Ignorance and Surprise

Inside Technology

edited by Wiebe E. Bijker, W. Bernard Carlson, and Trevor Pinch

Ignorance and Surprise

Science, Society, and Ecological Design

Matthias Gross

The MIT Press
Cambridge, Massachusetts
London, England

This book was set in Stone Serif and Stone Sans by Toppan Best-set Premedia Limited.

Library of Congress Cataloging-in-Publication Data

Gross, Matthias, 1969–
Ignorance and surprise : science, society, and ecological design / Matthias Gross.
    p. cm.—(Inside technology)
Includes bibliographical references and index.
ISBN 978-0-262-01348-2 (hc : alk. paper), 978-0-262-54398-9 (pb)
1. Restoration ecology—Social aspects. 2. Human ecology. 3. Science—Social aspects. 4. Knowledge, Sociology of. I. Title.
QH541.15.R45G75   2010
333.71′53—dc22

                                                              2009037131

Contents

List of Figures and Tables

Figures

Tables

Acknowledgments

I am very grateful for comments and inspiration from Kelly Moore, David Hess, and especially Wolfgang Krohn. Discussions with Alena Bleicher, Stefan Böschen, Scott Frickel, Holger Hoffmann-Riem, Christian Kuhlicke, and Peter Wehling have also greatly improved the book. Mark B. Brown, Arthur Mol, Otthein Rammstedt, and Michael Huber have read and constructively commented on one or more of the chapters. Paul H. Gobster from the United States Department of Agriculture's Forest Service has delivered invaluable materials and maps, as well as unpublished memos and minutes from meetings from the City of Chicago and the Chicago Park District's Office of Research and Planning. Research on the Chicago shoreline was begun in a research project on "Real-World Experiments" at Bielefeld University's Institute for Science and Technology Studies (IWT) funded by the Volkswagen Foundation. I would like to thank the Department of Sociology at Loyola University Chicago, my platform for conducting most of this research, for their support. Of the many people in Chicago who have helped me over the years, I want to mention Lynne Westphal, Kathy Dickhut, Buffy and Bill Jordan, Alannah Fitch, Chris Gent, and Suzanne Malec-McKenna. I am also thankful to the liberal work environment of the Helmholtz Centre for Environmental Research in Leipzig, where the book was finished. Without the early research on postmining areas by Sigrun Kabisch and Sabine Linke, I probably would not have learned about this fascinating field of landscape transformation.

Over the years, our little working group on mining impact assessments in Leipzig, consisting of Dagmar Haase, Martin Schultze, Katrin Wendt-Potthoff, and myself, has been a great source for getting a feeling for the surprising design of postmining lakes. I am also grateful to Andreas Berkner, Bernd Walther, Birgit Felinks, Bertram Schiffers, and Gerhard Strauch for

making papers from the Office for Mining (Büro für Bergbauangelegen-
heiten) and other internal materials available to me. Travel stipends to
different settings in North America and Germany from the American
Council on Germany (ACG) in New York, the German Research Foundation (DFG), and the Helmholtz Association are highly appreciated.

Ignorance and Surprise

1 Introduction: Brave the Unknown

Ignorance and surprise belong together. Surprises can make people aware of their own ignorance. A surprise is normally rendered surprising when it occurs unexpectedly and also runs counter to accepted knowledge. A surprise thus cannot be fully understood independently of a person's or a group's ignorance. In fact, novel things always include elements of surprise, uncertainty, and the unknown, all of which are located outside the sphere of prediction. Consequently, surprising events in research are often something to which scientists aspire in their activities since it means a window to new and unexpected knowledge. Scientific methods thus should allow researchers to surprise themselves as well as their peers. What is needed for doing so is an interruption of the continuum between accepted knowledge and future expectations. This interruption can be deemed an important foundation on which new findings are based. However, as Ludwik Fleck observed in 1935, "every new finding raises at least *one* new problem: namely an investigation of what has just been found" (1979, 51). New knowledge, in turn, allows for new options without delivering secure criteria for how these new options need to be handled.

The contemporary explosion of knowledge or the observation that our current age is the beginning of a knowledge society thus has a little remarked on corollary: new knowledge also means more ignorance. Thus, surprising events will occur more frequently and become more and more likely. If this is the case, handling ignorance and surprise becomes one of the distinctive features of decision making in contemporary society. The challenge in dealing with surprises lies in the fact that they lie beyond the spheres of probability and risk. This book examines the social prerequisites for surprises, the possibilities of their success, and their conceptual frameworks to better understand the handling of the "startling unexpectedness" that is inherent in all processes of innovation, change, and invention. To illuminate these larger theoretical arguments, the book scrutinizes the surprising effects that are fostered by human interventions into the environment and the way that surprises can be handled in the process of designing landscapes and the restoration of brownfields.

Landscapes, Precaution, and Experiment

Why use landscape design and ecological restoration as a touchstone? All over the world, regions exist where human activities have led to vast changes in the landscape—via industrial, military, and mining operations; urbanization and deurbanization; and the conversion of agricultural land or land for leisure activities. These human-made interventions have often triggered highly dynamic processes in landscapes and ecosystems and repercussions in society. To revitalize industrially altered landscapes, both policymakers and scientists frequently communicate that decisions have to be based on reliable scientific knowledge that should be free of uncertainty. Public debate on the limits of knowledge is often avoided on the assumption that it would undermine the public's confidence in scientific results. Therefore, both ignorance and its material consequences are usually externalized, and only known and observable "facts" are represented in policy and risk-assessment considerations. Ignorance thus becomes an "outlier" or a "black swan," to use Nassim Taleb's (2007) term, because it lies outside of regular and quantifiable events.

Not everything can be known, and absolute certainty is impossible—an inevitability that, as Weick and Sutcliffe (2007, 30) have noted, has culminated in "the hollow maxim 'Expect the unexpected.'" Furthermore, in everyday life most people know that many things happen to them unpredictably, but there still "is no conceptual language for discussing this thing that everyone knows," as Howard Becker (1994, 185) observed. Even so, when it comes to making decisions based on science and technology and communicating them to the public, undisputed scientific evidence with no side effects is demanded. Contrary to popular belief, there has never been a general belief in absolute truth and certainty among modern scientists and technologists (Moore 2008). The idea of certainty and truth through science, however, appears to be very much alive in official rhetoric. As researchers on modeling in global climate change (e.g., Shackley and Wynne 1996; Lahsen 2005) have shown, scientists often talk about uncertainty but use a rhetoric that suggests that this does not imply a serious challenge to the authority of science. Instead, it appears that certainty must be rooted in the rhetoric of scientific knowledge, whereas uncertainty and unpredictability indicates the presence of other or "non-scientific" sources.

To reassure a supposedly risk- and uncertainty-averse public, emphasis has usually been placed on further research on known uncertainties to create greater certainty and supposed reassurance that the risks at stake will be controlled. *Uncertainty* normally refers to a situation in which, given current knowledge, there are multiple possible future outcomes. To reduce uncertainty in planning and implementation, the traditional belief is that proper scientific results will almost automatically lead to policy and implementation. This linear idea has also been called the "cascade of uncertainty." Using the example of prediction in climate change, Steve Rayner (2000, 272–273) has argued that uncertainties in the science of basic earth-system processes are the basis for more uncertainties over emissions. This makes anticipation of impacts more difficult, which is followed by uncertainty about how groups of peoples will respond to such uncertain impacts. For many observers, cascading uncertainties appear to be a major hindrance to sound policymaking, and this cascade of uncertainty has led to a wait-and-see approach as well as a wait-until-more-science-is-available approach to climate policy. Thus, the funding priorities normally shift to the basic earth sciences to deliver knowledge that is more reliable. Instead of accepting uncertainty as a fact and then placing more emphasis on the possibilities of acting in spite of uncertainties, the least research effort is put into the social sciences and disciplines dealing with social and policy issues to absorb or effectively react to what cannot be avoided anyway. After all, coping with uncertainties and surprises is fundamentally different from predicting and preventing them. The challenge is how to knowingly and, increasingly, also publicly deal with what is not known without losing one's credibility or "scientific authority."

Some authors claim that the "precautionary principle" is aimed at cases where risks are poorly understood (see, e.g., Myers and Raffensperger 2005; Whiteside 2006). Originally, the precautionary principle was developed in reaction to the dominant regulatory standard, which requires affirmative evidence of harm before regulatory action can be taken. It was intended to promote government regulation of risky industrial processes by shifting the burden of proof from the proponents to their opponents (cf. Halfon 2007; O'Riordan and Cameron 1995). In practice, however, the precautionary principle has often been invoked to prevent government action in contexts of scientific uncertainty. It has often been interpreted as a call to delay action until more research becomes available—to wait until certainty about the absence of harm is proven and until preventive actions in face of uncertainties can be taken. Moreover, as Freudenburg, Gramling, and Davidson (2008, 23) explain, even if the precautionary principle were to be understood "as stating that scientific uncertainty should not be used as a basis for inaction . . . such a formulation does not spell out what *should* be taken as an appropriate basis for action." Precaution suggests what should not be done, not what should be done. Critics of the precautionary principle thus claim that it is contradictory and, if taken seriously, will block desirable changes and stop us from adopting better technologies. Cass Sunstein, perhaps the best-known critic of the precautionary principle, claims that the principle "purports to give guidance, but fails to do so, because it condemns the very steps it requires" (Sunstein 2005, 14).

I believe that both interpretations—the one that claims that precaution means paralysis and the one that says that precaution must be a key feature in regulatory politics—have not dealt seriously with the importance of ignorance and surprise. The critics ascribe a "better safe than sorry" attitude to the precautionary principle and recommend turning back to cost-and-benefit analyses and risk assessments based on known facts, thus ignoring the inevitability of uncertainty and ignorance. Proponents of the precautionary principle have not yet delivered any effective strategies for determining what exactly is to be done when decisions have to be made promptly and risk assessments or computer models cannot help in any meaningful way.

By way of building on these two opposing positions, an experimental approach will be developed in this book by reconstructing possibilities for acting in the face of (well defined) ignorance and outlining the necessary social and ecological capacities to cope with surprising events. The idea of experiment (from the Latin: *experiri*, "to try") will be of pivotal importance to link ignorance and surprise conceptually and to learn from and cope with the unexpected. The idea of an experiment as a trial or a venture into the unknown is a crucial momentum of any scientific enterprise, albeit experiments in the laboratory are characterized by detailed minutes and controlled procedures. Viewing ecological design processes outside the laboratory as experiments challenges premises of ecological predictability and certainty. After all, "experimentation is an effective strategy for sensing surprise" (Lee 1993, 58). Hans-Jörg Rheinberger (1997) has argued that what makes the physical, technical, and procedural basis for an experiment work is that it is deliberately arranged to generate surprises. Indeed, surprising effects of experiments can be seen as the motor force for producing new knowledge since surprises help scientists become aware of their own ignorance. Surprises are the impulse for unexpected knowledge. Put another way, similar to experiments in the laboratory, experimental activities in the real world can also bring surprises—but unlike in the laboratory they are often not welcomed.

In the following, a *surprising event* is understood as an occurrence that triggers awareness of one's own ignorance. It will be seen as a prerequisite to becoming knowledgeable about one's own ignorance (here called *non-knowledge*) as a basis for acting in the face of limited rationality and incomplete risk assessments. This calls for openness to surprises. Openness to surprises does not necessarily seek to prevent the occurrence of an event but accepts that the event will happen anyway. This approach becomes even more important if one acknowledges what ecologists have preached for a few decades—that there never was a balance of nature at any stage in its history and that to believe the contrary can at best be called romantic (cf. Botkin 1990; Pimm 1992; Krichler 2009). In other words, also from an ecological point of view, unexpected events can be called quite natural. Instead of waiting for the ultimate scientific truth about nature or a cost-benefit analysis about the risk involved in morphing the natural environment, a mindful openness to surprises can help turn unexpected events—including potentially threatening ones—into experimental practices or learning processes.

The challenge ahead is that new knowledge creates new options without delivering secure criteria for handling them. People may welcome the unexpected (since it creates opportunities for innovation), but they also seek to control, steer, or even reverse the surprising events. Understood this way, curiosity and the fostering of surprises enter a paradoxical relationship. They need to be both unleashed and controlled, if not at the same time then certainly in a well-organized and reflexive fashion. In this way, this study is a contribution to current debates on reflexive modernity (Beck, Giddens, and Lash 1994; Lash 1999), on ecological modernization (Huber 2004; Mol 2001; Spaargaren, Mol, and Buttel 2000), and the many small-scale experimental initiatives towards more sustainable communities. After all, as Gill Seyfang (2009, 185) writes, we need "policy changes to allow these initiatives the space to experiment and fail, free from regulatory controls which hamper evolution of ideas." In this way, the book is also a contribution to discussions on science and new forms of innovation in the knowledge society (Felt et al. 2007; Stehr 1994) and on science's changing ways of knowledge making (Funtowicz and Ravetz 1990; Nowotny, Scott, and Gibbons 2001). At times, ecological modernization and transformation will require that environmental goals be emancipated from purely economic values and incentives, which is something that the proposed call for a preparedness for surprising events takes as a pivotal element in the development of successful decision making directed by ignorance. Debates about knowledge societies and reflexive modernization generally circle around an expectation that in the twenty-first-century world the widening and deepening of scientific mechanisms of social control (and thus the level of uncertainty in the world) will increase rather than decrease. Furthermore, many relevant empirical studies in the so-called new political sociology of science (cf. Frickel and Moore 2006)—such as studies on knowledge gaps (Frickel 2008) or undone science (Hess 2007)—make a standard call for a new vocabulary to sort out different types of ignorance. I hope to contribute to this call. In this book, I regularly look at some of the leading points of these streams of thought.

Objectives

First, this book analyzes some classical and contemporary social science accounts of the role of ignorance and surprise in science and ecology and integrates them with the idea of experiment in society (chapters 2 and 3). In its second half, the book examines the role and importance of an accommodation to surprises and unintended turns in the dynamics of large-scale landscape change and restoration (chapters 4 and 5).

In chapter 2, the reader is introduced to ways that surprises can be theoretically framed as the driving force behind a notion of scientific activity and social life in general. Surprises give particular credit to the unsteerable dynamics of the natural world that cannot be reduced to social, institutional, and cultural issues. To focus on surprising events, a shift away from the focus on the classical institutional preparedness is needed, so that unexpected (or natural) phenomena involved can be reconsidered. In other words, rather than finding explanations mostly in social capacities, the focus on the surprising (and in this sense, nonsocial) elements points to the ignorance of human actors about the behavior of the natural or complex technical world. In this chapter, discussions of a "surprises-laden" development of modern society range from the classical writings of Georg Simmel to current debates on new modes of knowledge production. These concepts are scrutinized (using the coevolution of restoration as a public practice and an academic discipline) to understand strategies for dealing with unexpected events. The Chicago school of sociology's understanding of modern societies' experimental development is used to help explain surprises outside of the laboratory as a crucial part of handling unexpected events. To illustrate the normalcy of different types of surprises as they are perceived by different actors, the chapter finishes with a discussion of an ecological restoration strategy that includes unexpected elements in landscape development—the return of wolves to eastern Germany.

Since surprises can be seen as a prerequisite for making people aware of their own ignorance, chapter 3 frames different types of unknowns (nonknowledge, ignorance, and the like) as they appear in science and everyday practices. Using Simmel's perspective on nonknowledge (*Nichtwissen*) as a touchstone, the chapter develops a notion of how unexpected occurrences can be incorporated into an experimental model of scientific and technological development that includes the experimental handling of "surprises." In contrast to the work of many other writers on subjects dealing with ignorance, I develop a model to provide an understanding of how different forms of unknowns can be conceptually connected and how in the analysis of social processes these different unknowns can be chronologically related to one another. The advantages of such a dynamic model over ever-new taxonomies of types of ignorance are critical to grasping the many shadings of the unknown. The example I use to illustrate the development and chaining of different unknowns is malaria control. The model of different unknowns can be integrated into a broader notion of experiment as a novel form of dealing with new forms of knowledge production that are problem-focused, context-driven, interdisciplinary, and perhaps

even transdisciplinary. By so doing, I draw on insights from adaptive management and intelligent trial and error but also go beyond these approaches
by focusing on the experimental character and the (positive) tension
uncovered by the awareness of ignorance triggered through a surprising
occurrence.

Chapters 4 and 5 take up the concepts developed in chapters 2 and 3
to analyze in more detail how ignorance and surprise can successfully play
themselves out in ecological design projects. They look at ways of dealing
with surprises in unintended results in ecological restoration processes in
urban contexts and large-scale landscape transformation in former industrial areas. In some areas of landscape design and ecological restoration,
the acknowledgment of the unknown and the expectancy of surprising
events are accepted as a normal and sometimes even welcomed part of
decision-making processes. Thus, urban parks and postindustrial landscapes provide telling examples of the "normal complexity" of handling
ignorance and surprise in everyday life that will most likely gain in importance in many environmental fields on a global scale. Chapter 4 takes up
knowledge production in the field of restoration ecology and its shifting
boundaries as a proper science. The significance of the unexpected is discussed with the example of the morphing of Chicago's shoreline on Lake
Michigan from the nineteenth century to the present time. Of central
concern in this chapter is the appropriation of surprises so they can lead
to robust design strategies. Restoration ecology in a knowledge society does
not require simply more knowledge and new technologies. It is a type of
knowledge production that is permeated with surprising turns, which
unfold with the application of new knowledge. In this spirit, the design of
new landscapes is neither a linear process of master planning nor a process
of trial and error involving variation and selection but instead is a carefully
coordinated process of dealing with unexpected turns via experimental
practice.

In Chicago, robust strategies have been developed for dealing with more
or less rapid changes, but chapter 5 deals with the revitalization of former
surface brown coal-mining activities from the era of state socialism in
eastern Germany. The major focus here is the landscape that is south of
the city of Leipzig, one of the largest human-induced landscape changes
in European history. Comparing the experimental and postnormal character of the design of the shores of Lincoln Park in Chicago with the design

of new lakes emerging on the southern edges of Leipzig is enlightening for several reasons. Both regions are characterized by mainly state-funded and -initiated projects to restore and revitalize former degraded or less user-friendly areas. In both cases, the landscape is being "restored" without a historical reference point. In Chicago, a restoration in the strict sense would mean removing the landfill that since the 1860s has repeatedly pushed back the waters of Lake Michigan, and in Leipzig, it would mean returning the landscape to its appearance in the premining days from the same era. However, a major focus in the Leipzig case is on a reverse effect compared to the Chicago case. In Leipzig, early signs of success have become increasingly fragile due to lack of openness to surprises and acknowledgment of ignorance—that is, the transformation of ignorance into nonknowledge.

Finally, chapter 6 summarizes the book's main arguments but also assesses experimental practices in a broader debate on the contingency and reflexivity of current modernity. It links the overall argument of the book to recent debates on knowledge generation in what Ulrich Beck and others have called *second modernity*. In second modernity, human societies have begun to realize that not all risks can be controlled and that they must be coped with and incorporated into planning and development. Some of the challenges of implementing strategies that produce both scientifically reliable *and* socially accepted restoration and remediation processes are explored. Experimental practices as an accommodation toward robust research strategies are scrutinized to illustrate the benefits and limits of such an approach in contemporary knowledge societies. If we agree that surprises can be a good thing and yardsticks need to be developed over the course of experimentation, how can we avoid turning an experimental approach into a camouflage for failed projects?

A final word on terminology: I often use the term *ecological design* although the cases in this book deal with issues such as ecological restoration, the operation of earth-moving technologies, the recultivation of large landscapes, and the management of malaria. I hope that this term helps me avoid the heated debates on the proper definition of *ecological restoration* and its demarcations from other activities in the natural world. Consistent with current ideas in ecological restoration and related areas, the term *design* also points to a reversion to the attitude of traditional approaches to nature. It represents a step beyond the fatalism that is implicit in much of mainstream environmentalism, which at the core merely calls for a minimization of all human impacts on the natural world. If we define *design*, as some authors do, "as the intentional shaping of matter, energy, and process to meet a perceived need or desire," then design indeed is "a hinge that inevitably connects culture and nature through exchanges of materials, flows of energy, and choices of land use" (van der Ryn and Cowan 1996, 8). Understood this way, architects, concerned citizens, engineers, environmental scientists, and landscape architects are all designers. Design is not bound to a particular profession. *Ecological design* thus can be defined as any form of human intervention with the natural environment that attempts to improve natural conditions or reverse environmentally destructive impacts.

Given the increase in this kind of design activities, the surprising outcomes that emerge from these activities increasingly define the world in which we live. The question now arises how surprises and the not yet known can be incorporated into our theorizing about modern society. I hope that this book is a first step in that direction.

I   Concepts

2 Experiments and Surprises: Classical and Contemporary Perspectives

Planning is a factor in evolution that is characterized by random effects—that is, unforeseen, uncoordinated effects that to a certain extent act on evolution.
—Niklas Luhmann (2005, 191)

Classical social theorists of the late nineteenth and early twentieth centuries were interested in the modern increase in opportunities for social activities and in the ways that these activities bring unintended side effects. These included the liberating potential and acceleration of modern life, motorized transportation, new forms of communication and technologies, and almost any radical social change (cf. Calhoun 2007; Rosa 2005; Ross 2009). Some authors saw the increase of rationalization and "scientization" in the modern world as going hand in hand with an increase in unexpected and thus often derationalizing effects. This chapter tackles the surprising outcomes of rationalization by discussing some of the classical and contemporary sociological notions of the unexpected and of surprises, uncertainty, and the unintended consequences of modern society in general.

To continue this line of thought and to sharpen its terminological and theoretical basis, I scrutinize the general idea of knowledge production and its importance for our understanding of unintended consequences in modern societies. I begin with the claim that we are entering the age of the knowledge society and the ways that this relates to an actual increase in unknowns. This debate is accompanied with some interpretations of the work of the German philosopher and sociologist Georg Simmel (1858–1918) to develop a framework for a fundamental mode of understanding the unexpected in modern society. Simmel delivered some of the tools for developing a concept of progress as a path where unexpected occurrences (both positive and negative) can be positioned at the core of modern science and technology and are not understood as mere side effects. To illustrate some of the theoretical discussions, I highlight the centrality of surprises with observations of the practice of ecological restoration by relating them to current debates on new modes of research in contemporary knowledge societies. Of central concern is the observation that the practice of restoration and many other types of ecological design have always acknowledged that nature responds to human effort in surprising and unpredictable ways. Ecological restoration is related to new concepts that proclaim a shift from more discipline-based research (mode 1) to problem solving and more exploratory (and in this sense, experimental) forms of research in public (mode 2). In this new mode of knowledge production, planning becomes based increasingly on exploratory and experimental approaches. Here the traditional idea of an experiment gives way to a view that understands social development and evolution as an experimental performance that is deliberately arranged to foster surprises. Using this discussion, I develop different types of surprises by discussing the unexpected recurrence of wolves to eastern Germany.

Knowledge Societies and the Inevitability of Surprises

The specters of risk, danger, and hazards force the institutions that conduct risk analysis into a highly contradictory position in which their possibility cannot be ruled out, society cannot be insured against them, but potentially hazardous technologies must be implemented.[1] Alejandro Portes (2000) has argued that the unintended or the unexpected has been central to sociology ever since the early days of the discipline and that is was generally contrasted with "linear regularities," where consequences follow cumulatively from certain premises.[2] For Portes, sociology (unlike economics, psychology, and other social sciences) always seemed "to have a different, alternative vocation, defined by its sensitivity to the dialectics of things, unexpected turns of events, and the rise of alternative countervailing structures" (2–3). In fact, the sociologists' focus on uncertainty and risk taking and the economists' orthodox conception of rationality influenced many nineteenth-century economists to turn to sociology. One such an economist was Edward A. Ross (1866–1951), who later became one of the founders of American sociology. As a professor of economics, Ross objected to claims that the world is predictable and that consequences would follow from clear premises. For Ross, the focus shifted to the unintended and often quite unexpected consequences of goal-oriented activities in the modern world. In one of his last works on economics before he switched to sociology, he argued that uncertainty must be seen not as an anomaly but as normal and therefore as a basis for societal learning and new knowledge (Ross 1896, 112).[3]

Herbert Spencer, in his classic text on *Social Statics* (1850), also claimed that British acts of Parliament all arose out of the failure of previously passed acts (Spencer 1970; cf. Fine 2006, 4). In a similar vein, Max Weber, in his famous 1919 speech on "Politics as a Vocation," stated that "it is by all means true and a . . . basic fact of all history that the final result of political action often, not downright regularly, stands in a total inappropriate, often in a paradoxical relation to its original intention [*Sinn*]" (Weber 1992, 65).[4] However, unlike Ross or Weber, Spencer suggested that human societies were best left to their natural inclinations. His view of the appropriate role of political and reformist action excluded the state from meddling in industry and commerce, administering education, developing technology, or even taking action to improve sanitation. The view that the failures and drawbacks of purposeful development can indeed be seen as cornerstones of social progress appeared to be alien to Spencer's functionalism and at least unlikely to Weberian sociology.

A generation later, Robert K. Merton (1936) coined the term *unanticipated consequences*. Although Merton saw that unanticipated consequences could also have desirable effects (1936, 895), he developed this idea concerning scientific research and called it "serendipity"—that is, an anomalous finding that gives rise to a new theory (Merton 1968, 157–162; but see also Merton and Barber 2004). However, although serendipities are unplanned, a person or a group promotes them to gain positive results for modern science. A negative serendipity would be an oxymoron. Unlike in a public surprise, the underlying happening in the serendipity, as outlined by Merton, remains unobserved for members of the public.

Recent considerations in Germany have framed the unintended as *transintentional action* (Greshoff, Kneer, and Schimank 2003), departing from earlier debates at the German Sociological Society from the late 1970s, generally treating the unintended as a problem of social action—that is, as something not welcome and even as pure failure.[5] Other more recent sources on the unexpected and the importance of unintended consequences can be found in the work Aaron Wildavsky (1995). In Wildavsky's understanding, environmental and safety issues have become an area where greater knowledge has increased evidence of harm from technology (Wildavsky 1995, 433–447). For Wildavsky, a society without risk is a society that is unable to learn and therefore shows no progress. Society, however, shows a tendency toward risk aversion, which leads Wildavsky to the conclusion that progress is limited. What worries Wildavsky is that lower forms of knowledge as well as ignorance are devaluing expert knowledge.

All these observations appear to be especially crucial since in what today is understood as a knowledge society, scientific uncertainties via unexpected results are increasingly becoming part of the wider society. Originally, the term *knowledge society* was used to indicate the growing importance of expert knowledge as a structuring component in social relations and organization in developed countries (Bell 1973; Drucker 1973; Stehr 1994).[6] Generally, three phases are distinguished in the relationship between science and society from the seventeenth century to the present. First, the new science of the seventeenth century appeared on the scene with the promise to improve society, but these promises could not be fulfilled in their time (Krohn 1977). During the second half of the nineteenth century, the situation began to change when a type of science developed that was able to use its theoretical insight for applied settings. In this phase of modern science, the theoretical approaches and their idealized typologies appeared to be sufficient for practical purposes. The principles of electrical engineering, for instance, took off on their victory lap as an application of the theory of electrodynamics. Other areas (such as innovations in chemistry) also became increasingly important for everyday life through their development of a huge range of pharmaceutical products, cosmetics, and detergents. Science thus increasingly became a part of everyday life. The extreme diffusion of scientific expertise into society led to greater trust in science as a savior for social problems but also to an increasingly tighter coupling of science with various elements of society. The problems and risks of modern science and technology have been increasingly recognized.

Furthermore, science's expansion has repercussions not only for society but also for science itself (Weingart, Carrier, and Krohn 2007). Science was supposed to deliver answers to almost all problems, and expectations of safety and certainty through science were raised that could never be fulfilled. The expansion of science into virtually all fields of society meant that increasingly complex phenomena belonged to the realm of the scientist's explanation. In the field of technology implementation and in ecological projects, the many influences and the intricate reciprocal interactions in the field make it almost impossible to use the theoretical knowledge of the basic facts of a technology or an ecosystem to make any safe statement about its outcome. Taking these complexities seriously leads to the perspective that science, technology, and innovation cannot be solved by simply increasing expenditures on basic research and science. Thus, in many fields of research, science today has reached its limits for the solution of problems, and a certain public unease or mistrust in reactions to issues at the intersection of science and society appears to be on the rise. Quite often, the public is thought to fear science because scientific innovations entail too many risks (cf. Felt et al. 2007). Consequently, in a knowledge society, where scientific knowledge generation (among other issues) has become a major production resource, the changing expectations concerning science need to be taken seriously. A knowledge society thus should not merely be understood as an allusion to the increasing application of scientific knowledge and accordingly as a knowledge-based type of social order, as in Daniel Bell's classical work (1973), but should pivotally point to a society that is actively engaged in knowledge production. The insight that expert knowledge that is produced under idealized conditions in a laboratory has no counterpart in the real world has led to the hypothesis that increasingly the flexibility of the experimental method of the laboratory is extended to the general public. As Briggle (2008, 461) aptly describes it: "Yet as specialized knowledge expands, it is beginning to groan under the weight of its own bulk." Others have even claimed that society increasingly becomes a test field for scientific trials, which has led to the catchphrase of "society as laboratory" (Krohn and Weyer 1994).

Suggestions that knowledge in contemporary societies is, in one way or another, increasing in importance have been put forward for some time now.[7] However, these early attempts seem to differ from discussions that are more recent in that they often use *knowledge* synonymously with *information* or *news* and thus tackle problems of new information technologies or the enormous increase in and distribution of theoretical knowledge and information rather than the importance of normal side effects in modern knowledge production.

Some authors have claimed that the notion of the knowledge society simply presents an uncritical variation of 1950s modernization theory and its idea of a linear development to modernization along the Western model based on a scientification of society (e.g., Brint 2001b; Bittlingmayer 2005). Others question whether the idea of a society that is based on knowledge adds anything new and important to our understanding of contemporary society, since all societies in one way or the other have been knowledge societies. Some authors talk about the myth of a knowledge society (e.g., Kübler 2005). As Joseph Huber has pointed out, the notion of a postindustrial or knowledge society can be misleading because these labels merely point to "different aspects of industrial society progressing on its evolutive path" (Huber 2004, 9). Indeed, the new service industries and the production of knowledge discussed under the label of *knowledge society* are often even more resource- and energy-intensive than in any society before. Furthermore, the notion that new information technologies and the application of scientific knowledge are transforming social life in fundamental ways has long been seen as a general call of the Enlightenment and thus as nothing new that distinguished the late twentieth century from other eras (Fuller 2001; Rule and Besen 2008). However, unlike earlier authors such as Daniel Bell or Peter Drucker, the current debate on the rise of the knowledge society stresses that a pivotal feature of a twenty-first-century society is the increasing temporariness of scientific knowledge (e.g., Felt et al. 2007; Krohn 2001; Gross, Hoffmann-Riem, and Krohn 2003). At the same time, the demand for scientific relevance and the increasing pressure under which scientists operate call for new methodological and institutional tools that can capture the newly emerging social forces that act on scientific activities. The notion of a knowledge society thus also refers to the production and recombination of different types of expertise in settings that were formerly labeled nonscientific. Indeed, more knowledge can mean increased demands for further investment to achieve even more knowledge. As soon as more knowledge and brighter insights have been generated, the recognition of new issues that are unknown will not wait for long. In this book, the existence of a knowledge society indicates that the traditional juxtaposition between science and society—between knowledge production and knowledge application—has begun to change, perhaps even to merge. What could this mean in practice?

Generally, no one will object that scientifically produced knowledge in many cases is the superior type of knowledge. However, contrary to other scholars in science studies, Niklas Luhmann claims that an explicit orientation to scientific knowledge seems to be scarce in everyday life. For Luhmann, scientific findings do not help much in real-world situations. He compares science with the Christmas decorations in supermarkets and shop windows. The glamour and the abundance of goods and items might appear impressive at first sight, but "when looking for something in particular, one cannot find it, and when explicitly asking for it, the calculation of the shop owners is clear: we don't carry the items you are looking for" (Luhmann 1990, 641–642). However, is this really an apt comparison? Is the analogy between science in public and shopping before Christmas meaningful? Is scientific knowledge not indeed penetrating many areas in everyday life and thus transforming the risks of science into the wider society?

Many observers in the field of science and technology studies indeed believe so. Hence, to deliver an example of the development of ecology as a science and its emergence in its practical applications, I highlight the development of ecological restoration projects in North America. The challenges that this field has outlined for many areas of thought are further explored in current debates on change in science and in its application in a knowledge society—here, the debate between a proposed mode 1 and mode 2 in science. The thesis is that the growth in science and its shifts outside the clearly demarcated realms of the laboratory mean an increase in surprises. Therefore, when referring to knowledge society, we should focus both on the creation and the uses of knowledge and information and also on emerging areas of ignorance and surprise—understood as a normal characteristic and a major component of any human activity.

The Transdisciplinarity of Ecological Restoration

In North America, ecological restoration is a rapidly growing field of ecological research and practice, and it has attracted the attention of a large number of people interested in environmental issues since the 1990s. As developed on the tall-grass prairies of the North American Midwest over the past half century and more recently in other regions of the world, ecological restoration has reversed environmental damage and provided a context for negotiating the relationship between human society and other parts of nature (cf. Gross 2003a; Higgs 2003; Jordan 2003). To be sure, neither the idea nor the practice of restoration is entirely new. One could say that humans have practiced ecological restoration ever since farmers discovered shifting cultivation. However, in today's restoration projects, many practitioners employ strategies that have major parts of the work and planning undertaken by lay people and community organizations. In many areas, success in restoration and ecological design projects is measured in the context of practitioner skills and community goals and less so by the manifestation of underlying scientific principles. At the very least, the restoration of new natural areas will be beyond that of any single contributing discipline or interest group.

Although ecological restoration has come to play an important role in discussions relating to the environment and environmental policy over the past decade, different authors and groups understand the terms quite differently (cf. Gross 2002, 2006; Halle 2007; Higgs 2003). The most widely accepted definition is perhaps that *ecological restoration* is the active attempt to return an ecological system to a former condition following a period of alteration or disturbance through the reconstitution of processes, the reintroduction of species, and the removal or control of species that are inappropriate to the model system. The Society for Ecological Restoration International (SER) defines *ecological restoration* as "the process of assisting the recovery and management of ecological integrity. Ecological integrity includes a critical range of variability in biodiversity, ecological processes and structures, regional and historical context, and sustainable cultural practices."[8] Restoration is perhaps most readily understood as a form of environmental rehabilitation that is distinguished from other forms of rehabilitation by its commitment to the re-creation of all aspects of the model system, regardless of their value to humans. It furthermore emphasizes participatory processes aimed at a rewilding of the landscape that is being restored. This can also include unexpected and (as the example of wolves below shows) even unwanted elements. For Bill Jordan (2006, 26), ecological restoration should represent "an act of self-abnegation and deference to the ecosystem that takes the restorationist beyond economics to establish the deeper relationship with the ecosystem." Thus understood, the concept of ecological restoration promotes a special intimacy between nature and culture.

In Europe, the term *renaturing* has been in use instead of *restoration* (Gunkel 1996; Westphal, Gobster, and Gross 2010; Zerbe and Wiegleb 2009). It can be argued that the term *renature* moves beyond the debates about "restore to what time period?" and allows participants to envision the aspects of past, present, and future that they would like to be part of. It also treats humans, nonhumans, fire, wind, and so forth as participating in the shaping of this new space, thus avoiding another discussion on "What's natural and what's human?" Although the terminology is different, in many respects both renaturing and restoration activities have focused on the surprising aspects in the activities of designing or shaping a piece of land, addressed the negative influences of people on the "natural" landscape and the potential benefits, and the values attached to nature, so in this book I do not make a distinction between the two terms.[9]

Beginning in the late 1970s, the core strategy in restoration projects has been that hands-on practitioners who may have little or no formal training in academic ecology often achieve insights that contribute to and even challenge existing ideas about the ecology of the system being restored (Jordan 1994). For a long time, the diffusion of restoration knowledge thus occurred primarily as the original practitioners moved to new problem contexts rather than as results that were reported in professional journals or at academic conferences. Communication links were maintained through formal and informal channels. Through this kind of research, a good deal of often site-specific knowledge was lost after a few years. Until the mid-1980s, the state of science did not matter to the majority of ecological restorationists. Furthermore, because restoration was primarily undertaken by nonacademics and was practice oriented, the practitioners learned only as much as they needed to restore a system. They were motivated by the job to be done and the desire to care for and perhaps participate in the ecosystem rather than by the interests of any particular (academic) discipline. This led the academic-based scientist to become just one of a number of participants who were involved in restoration. Robert Cabin (2007b) even concludes that testing general scientific hypotheses and rigorous data in ecological restoration would be of limited use. Instead, he suggests that to accomplish ecological restoration as quickly and effectively as possible, participating groups should "be encouraged to perform informal, intelligent tinkering-type experiments, take pictures, and record some qualitative notes" (Cabin 2007b, 6).[10]

In this vein, many ecological restoration projects from their beginnings were practice-oriented and were carried out by amateurs who learned as much ecology as they needed for restoring ecosystems. This can be understood as a variation of what Harry Collins and Robert Evans (2007) labeled *interactional expertise*. Interactional expertise in ecological restoration can be developed through encounters with the natural world even though full scientific immersion is not reached. Indeed, much work in ecological restoration is based on a type of knowledge generation that has been labeled "discovery in the context of application" (Foray and Gibbons 1996). Restoration expertise thus is a form of expertise where the practitioners are not necessarily able to talk like academic ecologists but can actually do their ecology tests. In the long run, the practitioners' expertise can be turned into what Collins and Evans label *contributory expertise*, although the tacit knowledge-relating practice must be translated (for example, via focus-group meetings or other participatory processes) so that the practitioners' expertise contributes to the science. An example of such a process is discussed in chapter 4.

Thus, for some authors, restoration is understood as a new kind of science that connects knowledge production with its application and is transforming traditional disciplinary boundaries. Practitioners and observers of ecological restoration have applied various labels that indicate its novelty. Some talk of a new and sensitive natural science (Helford 2003), a community science that includes profound philosophical and cultural shifts (Higgs 2003), and an important field of the twenty-first century that integrates science, practice, and policy (Hobbs and Harris 2001). Others believe that ecological restoration resembles a merging of science and artistic creation (Turner 1987). Some try to connect restoration to nature conservation (Aronson, Milton, Blignaut, and Clewell 2006). Bill Jordan, who coined the term *restoration ecology* in the early 1980s, speaks of a new paradigm and a new communion with nature that shakes our traditional understanding of science (Jordan 2003).

By treating humans as a mature part of nature whose existence is now necessary for plants and animals to exist, ecological restoration as a nature-designing activity has little place for romantic notions of contemplation or humility. However, as much as this seems to be reasonable and has been accepted by most restoration ecologists, it remains unclear what restoration as the establishment of a "deeper relationship with the ecosystem" might mean in practice. As Bill Jordan has suggested, unexpected and unwanted elements (such as fire, mosquitoes, bears, and wolves) need to be included in the process of restoring ecosystems. Because humans are part of nature, they need to learn to act and move *in* and *with* the natural environment and as a *mature* part of it (cf. Drenthen, Keulartz, and Proctor 2009; Gross 2003a; O'Brien 2006). Exploring issues concerning performance in the writings of Hannah Arendt, Bronislaw Szerszynski (2003, 215) notes that today "knowing nature thus becomes much more provisional: nature comes to be known not as a fabricated object is known but as we know a participant in a dialogue." This questions the idea that nature is natural only when it is left untouched by humans. In this spirit, many restoration projects engage amateurs, laypersons, and professionals in restoring prairies, dunes, forests, and rivers in designing nature, with the expectation and maybe even the thrilling anticipation that nature will respond to human efforts in surprising and unpredictable ways.

With surprising unanimity, these practices and descriptions of practitioners veer toward the same conclusions as those that were arrived at in recent social analyses of science. These analyses proclaim a general change in science in the form of a new mode of knowledge production that is emerging and will affect our understanding of science in the future. A number of concepts reflect the changed relationships among science, practice, and the public. Some of those who proclaim a fundamental change in scientific knowledge and its mode of production also call for a transdisciplinary science or a mode 2 in knowledge production (e.g., Gibbons et al. 1994; Nowotny, Scott, and Gibbons 2001). A transdisciplinary context of application has "distinct theoretical structures, research methods and modes of practice . . . which may not be locatable in the prevailing disciplinary map" (Gibbons et al. 1994, 168). Transdisciplinary science is defined through its reference to and specific analysis of socially relevant problems. In a transdisciplinary perspective, actors need to be integrated into (ideally) all steps of the research process (Pohl and Hirsch Hadorn 2007; Scholz et al. 2006; Thompson Klein et al. 2001). Some of the authors who have been influenced by Helga Nowotny call for a socially robust science as a displacement of reliable science as it shifts from a mode 1 to a mode 2 form of science (Nowotny et al. 2001) will be reviewed here. In the proposed mode 2 of science, "transdisciplinarity is achieved by focusing on research problems as they emerge in contexts of application and where the heterogeneity of knowledge producers introduces additional criteria of assessment apart from scientific quality" (Nowotny et al. 2001, 223). Transdisciplinarity thus addresses issues of social, technical, and policy relevance where the primary goal is problem solving.

Others who have tried to frame the consequences of a new type of science talk about a "finalization in science" (cf. Schäfer 1983), the emergence of a "triple helix" among universities, industries, and governments (Etzkowitz 2008), or even a general declaration of a new age of postacademic (Ziman 1996) or postnormal (Funtowicz and Ravetz 1993) science.[11] In the rest of this book, I use the discussion of mode 2 mainly as an illustrative yardstick, not because the authors' conceptualization of mode 2 is more convincing or its empirical validity appears to be less doubtful than alternative diagnoses of changing science systems but simply because mode 2 is the most famous description of a transformation of contemporary science.[12] Indeed, mode 2 has fostered a debate that reaches the popular science papers as well as newspapers and weeklies. However, what unites all of the above-mentioned concepts with the mode 2 thesis is that these are all concepts that allude to changes in the organizational setting of scientific work, which are intended to achieve authority of scientific knowledge and especially the democratic embeddedness of such knowledge.

Ecological restoration, which started in the United States with a group of Midwestern practitioners, has also undergone dramatic growth as an academic discipline (Gross 2003a; Higgs 2003; Young 2000). Despite the self-descriptions of practitioners and the analyses of ecological restoration as a new form of science, a long-term perspective on the development of ecological restoration over the last quarter of a century shows a more unified picture. This coevolutionary picture is one of a reciprocally linked development between academic research and practical activities that at times include the ingredients of a new form of science (such as citizen involvement, lack of disciplinary or academic context, and the solution of defined social problems). This perspective on knowledge production shows how coping with ignorance and surprise has developed into a normal strategy that serves social needs and values (social robustness) as well as scientific reliability (epistemic robustness). Next, I illustrate the developments of some areas of ecological restoration by relating them to a propagated mode 2 for the dynamics of science and research in contemporary societies—that is, a shift from traditional discipline-based research (mode 1) to a problem-solving and transdisciplinary form of new science (mode 2).

New Modes of Experimental Knowledge Production

What exactly do terms such as *mode 1, mode 2,* and *knowledge production* imply? Most ecological restoration projects take place in areas that are populated by humans, so negotiations between heterogeneous actors and reactions to developments in different ecosystems become part of restoration work. Thus, many seemingly nonscientific factors are part of restoration practice, and questions arise about how ecological activities can be socially embedded to make their implementation more socially acceptable. Traditionally, engaging in scientific inquiry has required a shift away from socially relevant knowledge and toward knowledge that might not be understood or accepted by the society at large. Therefore, Nowotny and coworkers (Nowotny et al. 2001) call for a transgression or even blurring of the boundary between academic and nonacademic knowledge. For them, the only knowledge that can be called socially robust is the one that survives across different settings of many nonacademic groups with their own criteria of validity. Table 2.1 compares the two modes of knowledge production, mode 1 and mode 2, in terms of seven key traits. In the following chapters, I focus on the final three traits—means of evaluation, degree of validation, and planning—because they have received little attention in the secondary literature and yet appear to be crucial for our understanding of dealing with ignorance and surprise in settings of scientific work.

The new production of knowledge propagated by Gibbons et al. (1994) and Nowotny et al. (2001) has received a lot of attention, both positive and negative. Although the literature is extensive and there are many detailed criticisms of mode 2 attributes,[13] the main point of criticism has been the limited empirical validity of the mode 1 versus mode 2 thesis, especially the generality of the observation and its long-term historical perspective. This objection is understandable since Nowotny et al. have also claimed that we live in a mode 2 society in which "it has become

Table 2.1
Seven contrasting traits of mode 1 and mode 2 knowledge production extracted from Gibbons et al. (1994) and Nowotny et al. (2001). The final three traits have not been discussed much in the secondary literature.
Modes of Knowledge Production

Traits	Mode 1	Mode 2
Audience	Academic community	Wider society
Context	Disciplinary	Transdisciplinary
Organization	Hierarchical and institutional	Egalitarian
Top priority	Academic freedom	Social responsibility
Means of evaluation	Peer review and internal control	General social relevance and social robustness
Degree of validation	Scientific certainty	Uncertainty as part of the science
Planning	Long-term/linear	Exploratory/experimental

increasingly difficult to establish a clear demarcation between science and society" (2001, 47). This shift in the dynamics of science and research is between problem solving, which is carried out following the codes of practice relevant to a particular discipline—mode 1; and problem solving, which is organized around a particular application—mode 2. In the former, the context is defined in relation to the social norms that govern academic science. This, as the authors suggest, has tended to imply knowledge production that is carried out in the absence of some practical goal. In the new mode 2, knowledge results from a broader range of considerations. Such knowledge is intended to be useful to a certain group in society, and this imperative is present from the beginning. It is generated and sustained in the context of application and not developed first and then applied to that context. Knowledge thus produced is always produced under an aspect of continuous negotiation with all groups involved, including those that have nonscientific (that is, nondiscipline-specific or nonacademic) interests.

Framed in this way, the context of discovery (the production of knowledge) and the context of justification (the application of knowledge) coincide so that new knowledge can immediately enter into a practical implementation (Foray and Gibbons 1996). In the 1970s, authors such as Alvin Weinberg attempted to find terms for a new relationship between application and knowledge production, which he called "trans-science"

(Weinberg 1972). In Weinberg's idea, *transscience* is a movement beyond normal science and politics and toward answers to questions that lie outside the realm of science. However, the new forms of knowledge making can also be understood as processes of knitting together of what Sheila Jasanoff (1990, 76–83) calls *research science* and *regulatory science* or what Liora Salter (1988) calls *mandated science* in science policy. Most prominent in this respect is probably the difference between *research science* and *policy science* (Funtowicz and Ravetz 1993).

For those who propagate a new type of science, the pivotal point is that the world is witnessing a dramatic shift both in the institutional context of knowledge production and in the kind of knowledge that is being produced. Although Gibbons et al. are not overly clear on the issue, they sometimes contend that this process is well under way and is in fact "irreversible" (Gibbons et al. 1994, 11). Traditional research is mode 1, in which academics with separate roles develop knowledge in narrow fields of study and pass it on to practitioners. Understood in this way, mode 1 science did not interact with broad societal concerns. Gibbons et al. (1994, 140) believe that the traditional mode 1 type of research "will eventually become incorporated into mode 2 knowledge production and that the dynamics on which it rests will continue to unleash further institutional changes." In a mode 2 society, a transdisciplinary team includes practitioners and other interest groups that also produce new knowledge. The learning is immediate for all and is part of the discovery process. Solutions to the problems that are generated are beyond the resources of practitioners within a single discipline, and "knowledge will not be produced unless and until the interests of the various actors are included" (Gibbons et al. 1994, 4). Thus, criteria like aesthetic preferences become as important as peer review based on disciplinary considerations. In mode 2, the shape of the final solution will normally be beyond that of any single contributing discipline. That is why Gibbons et al. believe that it will be transdisciplinary. Transdisciplinary knowledge is to develop its own distinct theoretical structures, research methods, and modes of practice. As Gibbons et al. (1994, 9) write: "The loop from the context of application through transdisciplinarity, heterogeneity, and organizational diversity is closed by new adaptive and contextual forms of quality control. The result is a more socially accountable and reflexive mode of science." Since mode 2 is marked by the ever-closer interaction of knowledge production with a succession of problem contexts, the social organization of this kind of research takes place in egalitarian frameworks in constellations and teams that operate in informal social networks (cf. Ziman 1996). Nowotny et al. (2001, 179) furthermore claim that the "irreducible core of cognitive values and social practices, which once enabled good science to be distinguished from bad science (if not—quite—truth from untruth), has been both invaded—by forces once defined as extra-scientific—and dispersed, or distributed, across more, and more heterogeneous knowledge environments."

Contrary to mode 1, where research results are communicated through institutional channels, in mode 2 results are communicated to those who have directly participated in producing them. Although the authors often point to uncertainty as an important aspect for an understanding of a new mode in contemporary science, they do not elaborate on this very much. Indeed, if we connect the mode 2 thesis back to the notion of a knowledge society, then the increase of uncertainty in the way that knowledge is generated appears to be the most challenging point of contention. If knowledge is increasingly produced in the context of its application where the boundary conditions are hardly known and the risks are poorly understood, then surprising events—which trigger the ignorance of the actors who are involved—become perhaps the most important issue for science in a knowledge society. This has led to what Gibbons et al. (1994, 48) refer to as the volatile character of science in context, which does not allow for long-term planning: "Planning itself turns into an experiment and is to be seen as part of a longer-term societal experimental learning process." The real question is how we deal with these surprises as the result of new types of experimental knowledge production.

In short, although the viability of the mode 2 thesis is certainly limited, a few separate observations in the "mode 2 manifesto" (Hessels and Lente 2008) might well be worth further investigation. For instance, the lack of attention that is paid to experimental strategies becomes apparent when viewed from the background of the current debate on the "return of uncertainty," as Ulrich Beck called it, and the experimental character of all societal activities.[14] This is remarkable since authors such as Beck also have tinkered with the notion of experiment and "the world as laboratory" (Beck 1995, 122–127; Beck 1999, 60–61; cf. Latour 2004; Lemov 2005) but without developing the notion of experiment any further. That the notion of experiment does not and cannot model itself strictly on the natural sciences has been discussed by Ulrich Beck (1995, 125) in relation to the natural scientific experiments that have been carried out on humankind: "Natural science has thereby forfeited its exclusive right to judge what an experiment signifies." In other words, it is now the social sciences' turn to define what an experiment outside the laboratory signifies and what makes it useful. As a first step in reaction to this call, I develop a sociological notion of experiment that is built on different approaches ranging from current science studies to classical sociological writings. I then focus on the interdependency of the application and production of knowledge to outline a concept of experiment that models itself not on the ideals of the natural sciences but on the handling of surprises in society.

Public Experiments: Producing Surprise and (Occasional) Delight

The laboratory can be called the paradigmatic institution of modern science since it is the location where experiments are undertaken. Although Peter Galison (1997) has shown that what counts as an experiment and even who counts as an experimenter differs between different research fields and within particular disciplines, experiments are generally thought of as actions or operations that are set up to test a scientific hypothesis in settings that are detached from the rest of society. In the following, a notion of experiment is discussed whose connection to the laboratory ideal of experimentation is comprised of the idea of attempting to surprise an experimenter. In the strict definition of the controlled laboratory experiment, results obtained from an experimental sample are compared against a control sample where the independent variable is the only factor that varies systematically in the experiment. However, a control can be a most interesting part of the experiment since it often does surprise the experimenter (or anybody observing the experiment) by behaving differently from what was expected. As Bruno Latour (2004, 196) puts it, "A bad experiment is not one that fails, but one from which the researcher has drawn no lesson that will help prepare the next experiment. A good experiment is not one that offers some definite knowledge, but one that has allowed the researcher to trace the *critical path* along which it will be necessary to pass so that the following iteration will not be carried out in vain."

At the core, then, an experimenter is open to surprises but at the same time is eager to control the surprising event as a basis for learning. Unprecedented events in experimental processes, as Rheinberger (1997, 134) refers to them, "come as a surprise but nevertheless do not just happen. They are made to happen through the inner workings of the experimental machinery for making the future." As a scientific method, experimentation aims to manipulate the mechanisms and functions of the experimental system and to understand segments of reality represented by it (cf. Hacking 1983). However, taken out of the laboratory, experimentation implies a setup and a process without a fixed setting of an experimenter. The basic driving image behind the experimentalist idea in society is that every policy intervention is like or could be like an experiment. Real-world experiments explore new domains, generate learning through surprise, and in so doing help empower citizens to challenge accepted views. Yet the ability of such public experiments to generate learning through surprise is often overlooked, although an experimentalist approach can achieve the pragmatist vision of "mobilizing the public, revitalizing public discourse, and getting (citizens) personally involved in politics" (Shalin 1992, 245).[15]

Harry Collins (1988) as well as Steven Shapin and Simon Schaffer (1985) have discussed public experiments in the eighteenth century as scientific tests that were conducted in public to resolve issues such as doubts about the safety of technological innovations. Collins points to a paradox in the idea of public experiment, however, since so-called laypersons are expected to draw conclusions from experiments that normally require expert interpretation. Similar to the way that today's critics debate different expertises, Collins tries to resolve the paradox by distinguishing between "real experiments" and different forms of scientific display. However, in the twenty-first century, laypersons and amateurs again are routinely part of science (for example, in ecological restoration projects and related fields). Disease sufferers, as Kelly Moore (2006) has observed, can contest scientists' assertions about their treatments by basing assertions on bodily experiences to which scientists have no direct access.

Since the 1960s, Donald Campbell and his students have proposed large-scale societal experiments. For Campbell, the public is a place to test the efficacy of political programs (cf. Campbell 1988; Cook and Campbell 1979). According to Campbell, the political system is determined by the proponents' commitment to proposing success and the opponents' firmly embedded expectancy of failure. In this view, learning about reform politics must rest with the scientists. In its day, Campbell's well-intended approach was labeled technocratic. Taken full strength, some of the criticisms went, an "experimenting society" in Campbell's sense would mean no less than a technocratic colonization of the everyday world of potentially everybody (Fuller 2000, 28–30).

More systematically, authors such as Winfried Schulz (1970) place five core definitions of experiment in chronological order to indicate their respective major meanings in a certain era—experiment as a trial (trying things out by exploration), which can be found in all periods of human civilization; experiment as standardized scientific inquiry, which was used at the beginning of the modern era; experiment as a line of argumentation via a logical proof of causal relations (for example, the thought experiment);[16] experiment as material and machineries on the laboratory bench, as developed since the seventeenth century (cf. Greenwood, 1976, 48–71; Siebel 1965, 17–22); and experiment as a reformative change and renewal, which is what Schulz (1970, 22) implicitly sees as the most advanced form of experiment in contemporary society.[17] However, Schulz sees a problem with this idea since it would mingle science with practice. It is exactly the possibility of this intermingling between science and its practical implementation that is discussed in the following by looking closer into the history of sociology.

Concurrent with the idea of bringing (social) science and practice closer, in the 1890s the metaphor of the modern city as a laboratory was first used at the department of sociology at the University of Chicago. The perception of the city as a kind of laboratory and the study of human society as work in this laboratory was presented in the university's catalog in the academic year 1899–1900. It was claimed that "the city of Chicago is one of the most complete social laboratories in the world. . . . No city represents a wider variety of typical social problems than Chicago" (Tolman 1902, 116). Albion Small and George Vincent (1894), together with other sociologists of their day, believed that sociological investigation should be understood as taking place inside a social laboratory. This social laboratory, however, is a place where knowledge gain and practical work need to be combined. Thus, the significance of a scientific observation of society and the relevance of social reform went hand in hand. Beginning in the twentieth century, many different fields began to apply newly gained knowledge to society and design strategies that would feed knowledge directly back into society.[18]

In the 1920s, Robert E. Park (1864–1944), a student of Georg Simmel, took up the notion of experiment that had been embraced by the early founders of the discipline and marshaled the early Chicago ideas into a widely respected research program.[19] In Park's view, the modern city was a social laboratory where all plans that humans set out are tested within their own society so that the activities of the inhabitants can have unexpected consequences. In this context, Park often stressed the complexity and complication of social relations in modern societies, but at the same time, he believed that this offered new possibilities for its human settlers. In terms of Park's perceptions, the development of the city and of society at large can be understood to be associated with processes that "experimentally" result in a better understanding of how society works. Consequently, for Park, sociology is on its way to becoming "an experimental science." He explains that "experiments are going on in every field of social life, in industry, in politics, and in religion" (Park 1921, 177) and adds that in all these fields human beings are "*guided* by some implicit or explicit *theory* of the situation, but this theory is not often stated in the form of a hypothesis and subjected to a test of the negative instances" (ibid.). In other words, humans experiment in and with their own society. We could interpret Park here by viewing a hypothesis as being formulated by the individual's horizon of expectation. That is, when a surprising event occurs, an expectation serves as a (posthoc) hypothesis that has been tested.

Here Park is also elaborating an idea that society itself is operative in designing social experiments. In a similar way, John Dewey (1929) set out his pragmatist idea that the experimental methods of modern science provide a useful approach to the development of current social change. Dewey considered the relationship between knowledge and action by applying the methods of experimental sciences to societal learning. He also asserted that democracy provided citizens with an opportunity for maximum experimentation and personal growth. The perspective that meaning and reason were social in nature (that is, they required mutual cooperation and collaboration in their construction) led pragmatists such as Dewey to believe that an important part of the route to progressive social change lies in democratic deliberative approaches to addressing pressing social problems.[20]

In this spirit, Robert Park published his classic article "The City as a Social Laboratory" (1929). In it, the city represents modern society's most consistent and most successful attempt to remake the world in which people live and is the most prominent place for creating and supporting the experimental spirit. For Park, modern society has turned itself into a place that can be understood as a laboratory for investigating sociologists (cf. Gieryn 2006). Experiment takes place in society and is performed by society itself. The early Chicago sociologists and philosophers (such as John Dewey) were observing the increasingly experimental character of modern social life in general. For "experimentalists" like Park and Dewey, much of nature and social life was contingent, diverse, dynamic, or just unknown so that social constructions (including potentially competing, strategic ones) always needed to be checked for their utility.

Understood in this way, it could be argued that each new case of application of something is uniquely novel and therefore experimental in some respects. For most cases, this would be a trivial academic point, but for many of today's scientific and ecological applications and innovations, the artificial, controlled conditions of testing in laboratories or computer simulation models leave more and more salient questions about the application in the real world not just unanswered but unasked (cf. Wynne 2002). As discussed above, observers in science studies have long discussed that modern science tends to extend research processes and their related risks beyond the limits of the laboratory and directly into wider society. However, what so far has not been seen is public experimental strategies where the public initiates the experiment and does not merely react or adapt to it. The terms *public experiment* and *real-world experiment* thus can be used to point to the public character of experiments outside of the laboratory and to denote a change in the idea of the relationship between science and its publics in a knowledge society.

To conceptualize the difference between experimentation in the laboratory and real-world experimentation, Wolfgang Krohn (2007) suggests that both types should be compared to the nomothetic and idiographic approaches to reality that were introduced by the philosopher Wilhelm Windelband (1980).[21] Windelband saw nomothetic approaches to science as having the tendency to generalize from many cases to derive lawlike statements. Idiographic approaches, in contrast, highlight unique elements of single cases. Krohn therefore argued that nomothetic and ideographic approaches are both equally relevant for experimentation outside the laboratory. On the one hand, for research and implementation of novel technologies in the real world, "legal regulation requires taking into account the most recent state of reliable, certified, and licensed technology. . . . By and large this is knowledge of the nomothetic type" (Krohn 2007, 142). However, this new type of knowledge, which is generated mainly in laboratories of the natural sciences, needs to prove its robustness and usefulness in the real world, where it becomes, as Krohn suggests, idiographic. Instead of finding parameters for general (nomothetic) solutions, site-specific circumstances (such as geological conditions, human interests, or groundwater plumes) call for a strategy that keeps an installation going, even if surprises occur. Discussing the long-term study of waste-management policies and technologies, Krohn views the two sides (unlike in the original notion of Windelband) as complementary since "on the one hand, there is an increasing *contextualization* of research with a view to the planning and operating of installations and their natural and social environments. On the other hand, there is *decontextualization* of research in attempts to model and simulate the behavior of waste, to design new forms of construction, and to develop techniques of observation" (Krohn 2007, 143). The general observation has been that these types of experiments are characteristic of high technologies but that in many respects one can learn from them only by implementing them and trying them out. Thus, the problem that is posed by ecology-related risks cannot with certainty be solved on the basis of traditional experimental (field) methods alone, since human-nature interactions in the real world can cause highly unstable reactions—which are aptly captured by Charles Perrow's catch phrase of "normal accidents."[22] Understood this way, experimentation is a mechanism whose major aim is not to overcome or control unknowns but to live and blossom with them. Krohn (2007, 147) called this, in reference to Perrow, "normal experimentation." To conceptualize the idea of accepting surprising events as normalcy, I turn to some of the reflections of Georg Simmel and show that his ideas on the side effects of social processes can serve as a model that enriches our understanding of today's knowledge society.

Toward an Empirically Grounded Typology of Surprises

Simmel's idea of the trend of modern history includes a notion that social development is orchestrated by the interplay between two disparate parts.[23]

In Simmel's work, this phenomenon appears in his notion of a widening rift between subjective and objective culture. Objective culture has been created by people, is intended for people, but has attained an almost naturalistic form, which follows its own logic of development (Simmel 1998). The unexpected can also lead to positive outcomes and can have beneficial unintended consequences. Among the negative side effects, Simmel focuses on what he calls uncontrollable "objective" forces and sometimes a general "tragedy of culture." Nevertheless, some side effects (such as unexpected beauty in human-induced natural changes) are valued as positive.

Although Simmel saw the intensification of objective culture as a universal phenomenon in all cultural epochs, he felt that in modern societies and especially the modern metropolis it tends to become ever more encumbering. Understood in this way, Simmel's position at first sight appears similar to Marx's definition of *alienation*, Durkheim's *anomie*, and Weber's *rationalization*. However, Simmel further reflected on how modern society can *successfully* deal with these developments so that a causal relation between objective and subjective culture potentially can lead to a progressive development. Although Simmel identified a tragedy of culture, he also saw the potential to learn from the unexpected and tragic outcomes of human activities in modern culture so that in a new phase of "subjective culture," revisions and modifications to "objective" issues could be undertaken. To that end, Simmel delivered a fundamental mode for framing the unexpected in modern society as normalcy.

If the deliberate production of new knowledge for scientific innovation, cultural objects, social reform, or any other purpose always leads to new side effects or unintended consequences based on the creation of further nonknowledge (on the terminology, see the next chapter), then the idea of a knowledge society can help us deliver a more complete picture of social change. In other words, in the analysis and understanding of society, one has to acknowledge the presence of ignorance by considering unknown processes. This relates to Hannah Arendt's book *The Human Condition*, where the distinction between genuine action from mere behavior or habituated movements pointed to the "miraculous" initiatory quality that is inherent in all genuine human activities. In general, Hannah Arendt (1959, 176) reminds us: "It is in the nature of beginning that something new is started which cannot be expected from whatever may have happened before." Thus, knowledge production and its side effects via the acknowledgment of ignorance must be considered as constitutive for one another. Although Simmel is not explicitly referring to Immanuel Kant in this context, the acknowledgment of ignorance can be related to Kant's "transcendental categories" of understanding, which are categories that establish the limits of knowledge and lead to coincidental outcomes. Simmel's thinking on the development of modern science and technology resulted from his conjunction of relativist implications and his idea of the "objectification" of cultural achievements. This perspective departs from the linear idea that new technological innovations arise from the inner logic of science and technology, which regards new technologies as driving products. Science and technology scholars have detected representative cases from outside the realm of "science" that indicate how nonscientists reinvent and transform technological products, the results of which are fed back into the development of science and technology (cf. Eglash, Croissant, Di Chiro, and Fouché 2004).[24] Consequently, several streams in the sociology of technology and related fields have developed different models of networks of innovation, which suggest a recursive relation between scientific development and technical application.[25] In that sense, theorists of self-organization have talked about a "circular causality" (Küppers 1999) since a cause determined an effect, which in turn is manipulating its cause. In computer models, this feedback loop can run until the cause creates exactly the effect that its cause has reproduced (that is, a recursive process).

Some of the insights of the theories of self-organization can be used to deliver an understanding of unexpected occurrences (evaluated as positive or negative) by positioning them at the core of modern science and technology and not as indications of failures. As is shown in the preceding section, in Simmel's work subjective and objective cultures are two sides of modern cultural development. When human activity has produced a result, it may be the opposite of the one that was intended. Simmel's notion of objective culture illustrates that technical devices can encumber subjective culture. In an essay entitled "The Notion and Tragedy of Culture," Simmel (1998, 213) observed that "as soon as the human-made work is completed, it not only has an objective being and an individual existence independent of humans but also holds in its being . . . strengths and weaknesses, components and significances, that we are completely innocent of and that often take us by surprise."[26] Along with the etymology of the term *surprise*—which is derived from *sur* ("above") and *prendre* ("to take")—Simmel here points to the fact that although humans have invented something, their own work is able to surprise them through their own "objective" culture. More recently, Henry Petroski (2006) has argued that the success of new technologies has more often than not been built on the back of failure and not merely imitation and amplification of earlier technologies. As Petroski (2006, 94) says, "seemingly the best-designed engines can surprise their engineers." For Simmel, this character of what Arendt called the "startling unexpectedness," however, is inherent in the progress of all cultural processes, not only in the processes of engineering. The ecologist Crawford S. Holling (1986, 294) defined *surprise* as Simmel used it in the context of his relation between subjective and objective culture and points to Arendt's "startling unexpectedness": "Surprise concerns both the natural system and the people who seek to understand causes, to expect behaviors, and to achieve some defined purpose by action. Surprises occur when causes turn out to be sharply different than was conceived, when behaviors are profoundly unexpected, and when action produces a result opposite to that intended—in short, when perceived reality departs *qualitatively* from expectation."

Thus understood, something is surprising when a preexisting set of experiences and a horizon of expectation turn out to be inappropriate, since the real situation contradicts any anticipation. Building on the works of George Shackle and Joseph Schumpeter, the economist Neil Kay (1984, 69) defines *surprise* as follows: "A surprising event may be regarded as one whose occurrence was not anticipated, or which has been allocated such a low probability that the possibility of its occurrence was effectively discounted." This again suggests that there are known (imaginable) surprises as well as unknown surprises. Surprises can be perceived as unintended, imaginable, as well as unanticipated. Following Stephen Schneider (2001, 4671), an "imaginable surprise" can be defined as an event or a process that departs from the expectations of some definable community yet is a concept related to but distinct from risk and uncertainty. This distinction can be related to an earlier attempt by Shackle (1991, 422) to distinguish between "counter-expected events" and "unexpected events." The former can be related to Schneider's "imaginable surprise" where the opposite was expected and the surprising event was considered highly unlikely. The latter type of surprise is the one that has not been imagined because it did not enter into any reckoning.

All kinds of surprises are able to come out of the blue, although they can appear as anticipated post hoc. They are able to transcend the normal scheduling and routine mobilization of events. For Simmel, the question was important if the term *surprise* can be "freed from its psychological and sentimental meaning as a logical category for the relation between different contents, so that not every moment of life [*jeder seelische Augenblick*] would be surprising or novel" (Simmel 1922, 4). Simmel (1922, 8) concluded that the antagonism between moving back and forth between accepted knowledge to reach a goal in certainty and safety and knowledge about the unpredictability in all human action "needs to be viewed as a reality, that is almost like an object, a fact of life, that is part of everyday life's interdependences and its countless functions, which cannot be deduced from the conditions of knowledge but only the totality of life." Regarding the complexity and totality of life, Peter Timmerman (1986, 445) has said that "The heterogeneity of time and the complex physical relations between a system and its context necessarily involve the formation of surprises." Ecologist Harvey Brooks (1986) categorized different forms of surprises into unexpected discrete events, sharp breaks in long-term trends, and the sudden emergence into political consciousness of new information. All three types of Brooks's forms, which have been redescribed as "local surprise, cross-scale surprise, and true novelty" (Gunderson 2003, 36), are important for ecological design processes, but they call for different types of reaction. In the following, I refer mainly to local surprises, where an observer can register an unexpected event by attributing an expression of surprise to an actor or a group of actors.

As has been shown, for Simmel surprises can be perceived as negative but also as beneficial and pleasing. The notion of surprise as outlined by Simmel is elegant because it covers the idea of both unintended consequences and unanticipated consequences. A surprise, in Simmel's way of thinking, thus can be unintended but still anticipated.[27] If somebody knows that something is going to happen sometime somewhere but cannot stop it, he or she has no option for going ahead. Since the exact time, place, or ramification is not known, the event, when it comes, is still communicated as a surprise. In this sense, Aaron Wildavsky (1988, 93) detected two core categories of surprises. Quantitative or "expected" surprises occur when people know that a thing can happen but are surprised when it occurs in different amounts or patterns than is supposed. Qualitative, "true," or unexpected surprises are situations where the probability of consequences is not known. Wildavsky's first type of surprise would be an event where, for example, a probability is calculated via risk assessments, but people are still surprised that it happens at a certain time. The second category refers to a surprise where no risk assessment or probability calculation is possible or available in a certain setting. An example would be occurrences that scenarios never considered. In her study of the nuclear power industry, Constance Perin (2005) discusses an incident at the Davis-Besse power plant in Oak Harbor, Ohio. In March 2002, workers discovered that boric acid had eaten almost all the way through the reactor pressure vessel head. Before 2002, experts did not even think about the possibility that "nozzle leak deposits could eat into carbon steel of the reactor vessel, yet leaks had been a generic problem known to the industry since 1979" (Perin 2005, 5). There also can be surprises based on information that has been held back or has not been circulated enough. What can be concluded from the above is that "an event is not surprising or unsurprising in itself, but only in relation to a particular set of beliefs about how the world is; *of course*, a surprise is only a surprise if it is noticed by the holder of the beliefs that it contradicts" (Thompson 1986, 452). Climatologists might not be able to predict when and how strong a certain event will hit a certain region, but they can assure people that it will happen (Myers 1995). Other surprises, as we have seen, can be beyond the realm of expectation given existing knowledge. However, talking about an anticipatable surprise or even a "strategic surprise" (Floricel and Miller 2001) is a difficult task since it can be objected that events are always anticipated by some observers.

From a sociological perspective, it is thus not important to be able to distinguish a "real" or "total" surprise from one that has been anticipated. More important are how and why a surprise helps the actors to become aware of what they did not know (or what they had forgotten) and how and why they use this type of ignorance to move on in spite of it. One is tempted to paraphrase the famous Thomas theorem and say: "If human beings define a situation as surprising, then the surprise is real if it is taken as the source for further activities."[28] Alternatively, something is surprising when an observer registers a communication that utters perceived reality as different from a previous expectation and causally changes his or her subsequent behavior and activities since an event has run counter to accepted beliefs and knowledge. This is important since otherwise almost every type of change can be labeled surprising, a concern that also was shared by Simmel (1922, 4). For the empirical researcher, a surprise derives out of a difference between a reconstructable (that is, clearly uttered) expectation and one that is actually experienced.

Consider an example: Any ecological intervention is nearly always subject to complex processes of negotiation that can be controlled or directed only partially by scientists or some other specific social group. A targeted ecological intervention and a "design," no matter how varied the situations may be where it occurs (management of a contaminated region, design of a new landscape in a former mining area, or restoration of a eutrophic watercourse in the Alps), usually begins with an observation. The observers do not have to be scientists; they may just as well be walkers, joggers, or people without a PhD in soil physics who just happen to be driving past. The object of observation may be a lake, a fallow area of agricultural land, a backyard, or a river. Should the observations made run counter to expectations (the fish in the lake are swimming on their backs, the landscape is ablaze, a polar bear is romping around an urban backyard, or the color of the water in a river is red), then it is highly likely that subsequent communication about the impacts of this deviation or contradiction (which can be described as a surprise) will lead to a renegotiation of the stocks of knowledge about the segment of reality that has been observed. In the everyday world, experiential habits can assume the role of expectations. In scientific environments, this is usually done by suppositions formulated as hypotheses. However, a surprise cannot be registered in any meaningful way without an explicit range of expectations (or ones that are reconstructed in retrospect) that might be attributed to a particular actor group. Instead, it presupposes the existence of an observer who can establish a deviation from expectations. Seen in this light, there can be no surprise without anticipation (including a post hoc uttered nonanticipation) that can be attributed to a certain group or community. Without this, a surprise cannot be registered or observed in a sociologically meaningful sense. This is not a trivial point. Peter Baehr (2002, 822) talks of unprecedented events that are surprising "if it is impossible to identify an earlier event that is *sufficiently* comparable to the more recent event in *relevant* respects." However, this means that the observer of such an event would have to decide what "sufficiently" and "relevant" mean in a certain case to prove a fundamental dissimilarity compared with events that have happened before. Instead, it seems more practical to take the observed actors' self-descriptions about their reaction seriously and decide if an event was perceived as a surprise based on their members' own practices and descriptions of a situation. Furthermore, in this context, a surprise needs to be seen as an impulse for becoming aware of one's ignorance so that future behavior is changed to generate new knowledge (figure 2.1). A surprising visit of an old friend on a Sunday afternoon, although unexpected and pleasant, is not the type of surprise that will change one's future behavior qualitatively. Such a visit normally does not change one's overall worldview or foster new knowledge production. It does not stimulate what Argyris and Schön (1978) call double-loop learning or what Hedberg (1981) refers to as "turnaround learning." Both concepts allude to a process of rethinking the assumptions, routines, standards, and decisions within an individual, a group, or an organization. The unexpected visit of an old friend can thus be framed as single-loop learning, since the reaction to the visitor normally does not call for a change in one's knowledge sets and fundamental belief systems.[29] Instead, it calls for adjustments inside a person's or a group's worldview or paradigm that belongs to routines and standard reactions to surprising visitors (such as keep some extra beers in the refridgerator).

Furthermore, even unanticipated consequences (that is, total surprises) can be pleasant and beneficial. They can be based on unintended events

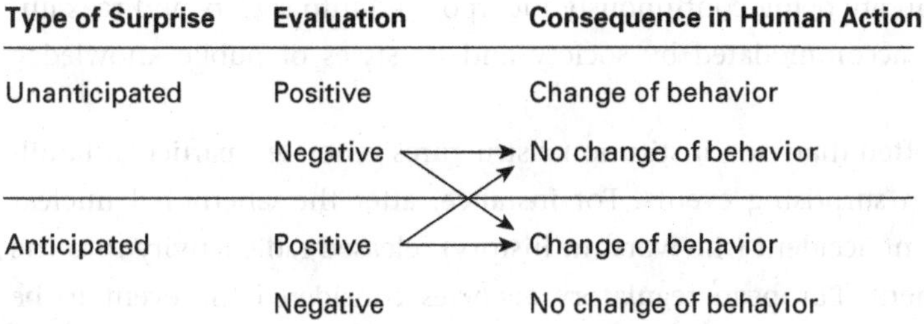

Type of Surprise	Evaluation	Consequence in Human Action
Unanticipated	Positive	Change of behavior
	Negative	No change of behavior
Anticipated	Positive	Change of behavior
	Negative	No change of behavior

Figure 2.1
Types of surprises and their evaluation

and can be anticipated and unanticipated, pleasant and desirable, as well as unpleasant and undesirable. Surprises thus can be understood as a basis for opportunity since they allow learning from the outcome of a surprise. In this context, learning can be understood as a communicative practice that produces new knowledge as well as new ignorance. Indeed, learning in societal contexts is different from learning in small groups and in individuals. It includes goals, hypotheses, expectations, and often long and laborious processes of decision making before an activity can be initiated. For the purposes of this book, *learning* is understood as "societal learning" to differentiate it from individualist "social learning." Societal learning can be understood as the collective action and reflection that occurs among different groups as they work to improve the management of human and natural interactions (cf. Keen, Brown, and Dyball 2005). As is discussed further in the context of Simmel's notion of nonknowledge, a key challenge to understanding societal learning is the importance of trust and collaboration among concerned citizens, researchers, planners, and engineers (Dodgson 1993; Gubbins and MacCurtain 2008). Furthermore, much of collective learning must also include openness to cooperation and, perhaps even more important, the "pressure to learn" (Borowski and Pahl-Wostl 2008) based on a common goal within a multistakeholder context.

When talking about learning, I also refer to what has come to be known as *civic epistemologies*—that is, the tacit rules by which collective knowledge is produced and validated (cf. Iles 2007; Jasanoff 2005; Miller 2005). In this way, the social practices of everyday life provide the basis for knowledge production and learning in public (Brand 2010; Reckwitz 2002; Schatzki 1996).[30] Thus understood, a successful development of learning can be maintained only when the unexpected side effects of modern knowledge production are being continuously incorporated into society and reexamined and accommodated by society and its styles of public knowledge making.

More often than not, institutional structures generate a particular familiarity with surprising events. For instance, after the Chernobyl nuclear power plant accident (the worst in history) released radioactivity into the environment, European regulatory agencies considered the event to be "noncritical." Even on October 15, 1986, a full six months after the Chernobyl accident, the German Reactor Safety Commission (RSK) stated that with this accident, "no new phenomena occurred. No new or surprising processes have been observed" (as quoted in Huber 2008, 121). This example shows that although scientific actors might not have been surprised at the scientific level, at the public level of the media and the policymakers, the event was clearly communicated as a surprise that could be called anticipated and led the public to question the authority and value of certain types of scientific expertise in public discourse.

Simmel was interested in detecting the possibilities and capacity of subjective culture to use, absorb, and transform elements of objective culture. Following a state of "objectivation" of subjective culture, Simmel assumed that both positively and negatively evaluated surprises need to be "resubjected" or "resubjectivated" (Simmel 1998, 213) so that they merge into new inventions and cultural achievements. For Simmel, the conversion of an invention into something completely different is sometimes welcomed. He saw negative side effects as rooted in processes of rationality and differentiation but also as possible precursors to a new uprising—one could say as a new cycle of practice and potentially learning from the old developments. Following Simmel, scientific knowledge production and its side effects in the context of application must be considered as constitutive for one another. Since today it seems generally accepted that knowledge expands and science grows at an ever-increasing pace, it also can be followed that anomalies are, as Bauer (2001, 461) put it, "a regular and necessary part of the scientific revolutions that mark the progress of science, it follows that they will crop up more frequently." This crucial insight is important for understanding the scientific production of nonknowledge in a knowledge society.

Before turning to an in-depth discussion of ignorance in knowledge production, let us consider an example of the unexpected return of the gray wolf and the attempts to control an animal that was thought to be extinct in most central European countries. Given that Germany, one of the world's leading industrial nations with 80 million inhabitants, is about as large as the state of Montana in the United States, the return of wolves after a 150-year absence is a surprising comeback. The comeback of wolves to eastern Germany has happened unexpectedly, and the animals are often unwelcome by many of the human inhabitants of the region along the German-Polish border.

New Nature, New Surprises: Return of an Extinct Carnivore

Wildlife such as bears, lynxes, and wolves are unwelcome by most farmers but welcomed by environmentalists, some government officials, and many segments of the public. The eastern German region of Lusatia, which includes parts of the states of Brandenburg and Saxony, has recently received a lot of attention in the German media because of the return of the archetype of the unwanted and dangerous animal—the wolf. Wolves died out in most parts of Europe in the eighteenth and nineteenth centuries. They were first spotted again in Lusatia in 1998 after they migrated back from western Poland (cf. Stoepel 2004). Today, wolves are one of the few existing animal species that survived the last ice age around 300,000 years ago. Over the years, gray wolves have been highly adaptable to different types of environments ranging from forests, deserts, mountains, tundra, taiga, grasslands, and, increasingly, urban areas. In eastern Germany, packs of wolves have unexpectedly (and against the wishes of some human inhabitants) returned to the region along the German-Polish border.

Viewed in simple terms, the growth of natural areas is a natural consequence of population decline and deindustrialization: less industry and fewer people mean more nature and thus more space for dangerous species. In the real world, the situation is more complicated. Despite a human population decline in these regions of Germany since 1990 due to outmigration and low birth rates and unlike other areas where wolves have reappeared such as Yellowstone Park (McNamee 1997), Lusatia is still a relatively densely populated area with currently 107 inhabitants per square kilometer. Hence, the returning wolves did not arrive in a depopulated region where the main problem is livestock lost to wolves.

Although wolves do not depend on depopulated open areas and can adapt to diverse environments, many restored forests, former military areas, and rehabilitated fauna of former mining pits nevertheless have delivered increasingly favorable living conditions for the returning animals. Since 1998, at least two packs have settled in the area, and the research group Lupus of the Biological Research Station in Spreewitz has documented over 250 peaceful encounters between people and wolves in the region. Nevertheless, wolves are not always welcome by the human inhabitants of the region. Hunters complain that wolves cause economic damage, and hunters' associations are skeptical about the development in general.

As a reaction to these complaints, shepherds, farmers, and ordinary citizens in the German state of Saxony have officially become entitled to compensation payments when their livestock is harmed.[31] This development has actually fostered the acceptance of wolves, which are increasingly considered as a normal part of the natural world near human settlements.

The original appearance of wolves to many people in eastern Germany after more than 100 years has been an unanticipated surprise, since wolves in civilized and densely populated regions appeared to be beyond any type of reckoning. However, the appearance of any new wolf continues to be communicated as surprising (for instance, in the local and national media) but also as somewhat anticipated. This example illustrates the tension in experimental strategies between an openness to surprises (let the wolves return if they "want" to) and the control of the "unexpected." In the wolf case, the control of the unexpected has led people to attach tracking devices to as many wolves as possible to be able to monitor where they "unexpectedly" move. When a wolf is spotted, it can be anaesthetized and be fitted with collars with tracking equipment attached to them. Thus understood, the unexpected behavior of the wolf becomes a controllable normalcy. Whether the appearance of wolves is considered surprising depends on the interests of certain social groups and their previous sets of knowledge and perhaps not so much as whether the wolves were anticipated by someone. When viewed from the perspective of wolf researchers and other proponents of wolves, the anticipated surprise is welcome, but from the perspective of hunters associations and farmers, it is often viewed as negative (see figure 2.2).

In an overall evaluation, Reinhardt and Kluth (2007, 65–70) showed in a social survey that (despite the still popular story of Little Red Riding Hood) most Germans today have a positive general attitude toward wolves, which supports the assumption that surprising sights or traces of wolves will be rendered positive. Hunters' associations and the government call for a "reasonable management" of wolf packs. Although Wolfgang Bethe, president of the state hunters' association, called the return of the wolf an enrichment for nature, he also stressed that wolves compete with hunters, since hunters have to hunt according to animal-protection laws, whereas wolfs hunt "as they like," as reported by the *Berliner Morgenpost* ("Jäger diskutieren über Rückkehr der Wölfe" 2007). Notifications of losses from farmers and citizens at the Biological Research Station have increased

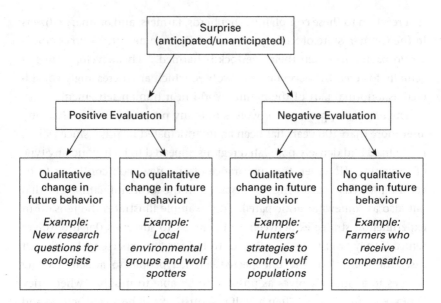

Figure 2.2
Types of surprises, their evaluation, and possible consequences

considerably since 2005. Thus, establishing reproducing wolf populations has been restricted to two small regions bordering Poland. One of the guidelines for the coordination between local authorities in Germany and Poland has become (contrary to the ideal in ecological restoration) that contact between humans and wolves should be kept to a minimum. Among the complex and problematic issues of monitoring the wolf populations is deciding when a wolf has behaved badly and needs to be shot. At what point has the unwanted wolf overstayed its welcome?

There are also attempts to engage the public in the process by encouraging people to observe and report wolf behavior. In Upper Lusatia, cyclists are invited to ride their bikes along the Wolves' Cycle Path, which is marked with the logo of wolf tracks to follow known wolf routes, read their tracks, or even spot and observe wild wolves. The local media support the monitoring process with reminders of help lines and the location of information offices (Kluth and Reinhardt 2005). They have involved the public in these projects to try to ensure that they will become important stakeholders in the process. Ordinary citizens can develop important knowledge about the wolves in their area and can share their knowledge with academic scientists by reporting their observations to one of the research stations. This is a way to cope with the surprising return of wolves, to turn concerned citizens into active participants, and to treat the general public as a valued member of ecological projects. Although this appears to be an indicator of a mode 2 type of science, it nevertheless leads to the development of academically oriented research on wolves in a mode 1 fashion. In addition, the unexpected return of wolves has considerably changed existing assumptions about human settlements in the landscape of an industrial country. It has also changed discourse in the national media on wilderness and on what is natural. In short, the reintroduction of wolves is very different from many of the museum-type restorations that are prominent in many North American contexts (such as arboretums).[32] If surprising events, as outlined by Simmel and as exemplified by the reintroduction of wolves to eastern Germany, can help people become aware of their own ignorance, then we need to ask whether we are able to integrate this insight into a broad conceptual model of modern societies.

In the next chapter, I will further explore how to analyze the unknown and connect this to this discussion on surprise. How can the awareness of our own ignorance via a surprising event be turned into a specific type of ignorance? I try to answer this question by integrating connotations of sociological terminologies that try to grasp the unknown and by outlining the dynamic and recursive relations of these types of knowledge and the way they can change over time. I illustrate this with the example of the development of malaria control. To knit together the findings on different notions of surprise and a model to grasp the dynamics of different types of unknowns, I round off the next chapter by outlining a working model of an experimental integration of ignorance and surprise that serves as the foundation for the cases discussed in the rest of the book.

3 Knowledge Production and the Recurrence of Ignorance

The fact that we can know our knowledge as well as our nonknowledge, and that we can also know this encompassing knowledge and so on into the potentially infinite—this is the actual infinitude of life on the level of the spirit.

—Georg Simmel ([1918] 1999)

Until the 1960s, sociologists of science normally separated any definition of *knowledge* from *belief*. Belief has been something less than knowledge since it lacked warrant or justification. Knowledge in this way has been understood as derived from an encounter of a person or a group with reality so that logic, rationality, and evidence determine what knowledge is. Since the establishment of the social studies on science and technology in the 1970s, this view has been disputed, and today both knowledge and belief are normally related to any type of social order. In contemporary social thought, *knowledge* in its broadest sense is therefore mostly referred to as a (justified) belief that is connected to purpose or use and is associated with intentionality. In the majority of cases, knowledge is perceived as true knowledge or at least as knowledge that someone holds to be true. Although it would certainly make sense to talk about "false knowledge" of a person or a group, in everyday life one hardly finds a difference between truth and knowledge. If something is known, it is also taken to be true. Thus understood, what a person knows must be true knowledge; otherwise, it would be not knowledge but a deception or fraud. Since most debates in science are fought over the level of certainty to assign claims, truth points to the scrutiny of a knowledge claim (cf. Luhmann 1990, 167; Pinch 1981). Reminiscent of Francis Bacon's observation that knowledge is power, Gilbert Ryle (1949) argued that knowledge is a capacity to perform in particular circumstances.[1] This view holds that knowledge is dependent and embedded in a context of specific social conditions.

This chapter continues the discussion of the ways that surprises and ignorance are related by clarifying different terms, concepts, and shadings of the other side of knowledge, the unknown. To understand how different fields of ecological design deal with surprises, in the following section I review some major streams in debates discussing the unknown. I suggest ways that some of these usages and connotations of concepts can be bundled and examine the dynamic relationships among different types of unknowns and their changes over time. The discussion focuses on the development of malaria control from the 1940s to the current date. Based on these reconstructions, I suggest a simple typology of notions of the unknown and their (sometimes) recursive relationships to each other in which existing usages of knowledge about the unknown can be situated without excluding each other. Because the experimental production of new knowledge and new inventions always creates new ignorance, ever-new perceptions of the unknown lead to further unintended consequences. This is the pivotal point for a knowledge society. The relationship between ignorance and surprise is approached in the context of real-world or public experiments so that a framework integrates ignorance and surprise to allow the study of different cases of ecological design and landscape development.

Knowledge in a Sea of Ignorance

Knowledge has always had a major function in social life. In the first half of the twentieth century, the sociology of knowledge began as the study of the social origins of knowledge and of its effects on social development. In the early version of the sociology of knowledge in the tradition of Max Scheler (1926), the following categories of knowledge distinguished the knowledge of salvation, cultural knowledge, and knowledge that produces effects. In 1958, Michael Polanyi distinguished between "explicit knowledge" (which can be articulated in formal language and transmitted among individuals) and "tacit knowledge" (which is embedded in individual experience and involves personal belief, faith, and morals). He also implied that all types of knowledge are constituted in practice and that knowledge cannot be a given capability but needs to be grasped as an ongoing process.

Thus, knowledge does not exist "out there" but is continually enacted through individual and collective actions and practices.[2]

By the late twentieth century, however, the view that knowledge as capacity for action depends on specific social conditions was not very helpful since it did not tell us much about the changes of knowledge in society today. Some authors thus introduced the notion of *knowledge worker* or the category of *knowledge-intensive work* (Drucker 1973; Wuthnow and Shrum 1983; Willke 2002) to denote a new class of knowledge producers that includes scientists, researchers, and managers. In general, a knowledge worker is a person who develops and uses knowledge in the workplace. Helmut Willke defined "knowledge work" in contemporary knowledge societies as characterized by permanent revision processes since knowledge can never be final. Furthermore, Willke (2002), similar to Stehr (2006, 305), observed that knowledge is decreasingly seen as a resource for speaking truth and increasingly as simply a resource for action. Finally, this all leads to the insight that with more knowledge, its counterpart—nonknowledge—will also rise.

In addition to attempts to define and categorize different forms of knowledge, the focus on unknown processes and variables has become increasingly important in theorizing about society and the production of knowledge in the twentieth century. In writings on reflexive modernity (Beck 1996), the risk or knowledge society (Knorr Cetina 2007), and the general crisis of knowledge in current intellectual thought (Wallerstein 2004), terms such as *ignorance, nonknowledge,* or *negative knowledge* are used to denote that something can and indeed must be known about what is still unknown. This debate about nonknowledge and ignorance goes back at least to Socrates' insistence that his wisdom lay in knowing what he did not know, which occasionally is referred to as *nonknowledge* (Jaspers 1951) but mostly as *ignorance*. The terms that are used in today's debates are increasingly connotated with different meanings that are sometimes antithetic to one another in their implications. Some of the current problems can be traced back to translations of the term *Nichtwissen* from the German into the English.

More generally, attempts to grasp notions such as ignorance are spurred by a well-known paradox: whenever new knowledge arises, the amount of nonknowledge that is perceived can increase proportionally since every set of newly generated knowledge can open up a wider horizon of what is not known. Alternatively, as Georg Simmel (1999, 303) put it: "Against the increasingly growing and unforeseeable progress of our knowledge, we should not overlook that in the end many things that we thought to own as certain knowledge can degenerate to uncertainty and become recognized as error." When discussing the history of an early version of the first law of thermodynamics, Ludwik Fleck (1979, 20) wrote "that there is probably no such thing as complete error or complete truth. Sooner or later a modification of the law of conservation of energy will prove necessary, and then we will perhaps be obliged to fall back on an abandoned 'error.'" Indeed, the modern idea of science as a means for turning uncertainty into certainty instead has more often led to more knowledge about what is unknown and perhaps cannot be known. What seems to emerge from this process today is an increasing awareness of the relation between new knowledge production and perceived ignorance or a deficit in knowledge. In discussions of complex behavior, complexity is rendered as a source of uncertainty and surprise not because of a lack of proper knowledge but because of a loss of predictability.[3]

Blaise Pascal (1623–1662) described knowledge as a growing ball, which moves in a sea of ignorance. When knowledge is thus imagined as a ball, a growing ball means that the surface that is bordering ignorance also grows. Jürgen Mittelstrass (1996, 34) offers two possible interpretations for Pascal's metaphor. In the first interpretation, the radius of the ball represents knowledge. As the ball grows, its surface area grows faster than the length of its radius. If this interpretation is correct, then ignorance grows faster than knowledge, and scientific research produces ignorance faster than it produces knowledge. In the second, more optimistic interpretation, knowledge is not represented by the radius but by the ball's volume. If the ball grows, then its volume grows faster than its surface area. In this case, research produces new ignorance, but knowledge grows faster than nonknowledge. Both interpretations suggest that growing knowledge increases the realm of ignorance.

More knowledge means more surprises, more awareness of ignorance, and new ways of coping with the unknown—that is, treating surprises as the norm. As a first step to coming to terms with the importance of different types of the unknown, I clarify some concepts and terms that deal with the unknown and suggest ways in which they relate to each other.

Nescience, Ignorance, and Nonknowledge

If the observers of the current knowledge society are correct, then an orderly and normal part of scientific activities will be dealing with ignorance. In his now classic book *Ignorance and Uncertainty*, Michael Smithson observed that in the second half of the twentieth century we "have seen a flurry of new perspective on uncertainty and ignorance whose magnitude arguably eclipses anything since the decade of 1660 which saw the emergence of modern probability theory" (Smithson 1989, 3). Some four years later, Smithson (1993, 133) wrote that it "is still not entirely respectable to write about 'ignorance.'"[4] More recently, Peter Wehling (2006, 18) observed that discussions of "ignorance, although still marginal, have developed into an increasingly relevant topic in sociology." Although today the importance of dealing conceptually with ignorance appears to be accepted, the terminology for grasping the unknown is still sketchy, and theories of ignorance have not yet been elaborated. Indeed, at any conference where a presentation is given on knowledge gaps, ignorance, or nonknowledge, the debate afterward circles around proper definitions, new taxonomies, or lengthy new terms. Because of this, authors tend to give extended definitions of terms every single time they mention a social phenomenon dealing with the unknown. This has lead to an enormous increase in the number of adjectives that are placed in front of a noun. Kerwin's (1993) "unknown unknowns," Smithson's (1989) "meta-ignorance," or Ravetz's (1993) "ignorance of ignorance" or "ignorance-squared" certainly can be handled easily enough. But some carefully constructed terms—such as "unspecified known ignorance" versus "specified known ignorance" (Böschen and Wehling 2004) and combinations like "openly reducible personal ignorance" (Faber, Manstetten, and Proops 1992)—rarely lead to a clarification because they are often used counterintuitively and are only partially grounded in concrete examples. Some of the taxonomies seem to be largely terminology-driven and to have few links to concrete examples or data.

Furthermore, the known taxonomies of ignorance have a linearity that does not allow connections between, for instance, "unknown unknowns" and "specified nonknowledge" to be identified, although the empirical reality usually suggests exactly this kind of connection. In short, simplification is needed. I suggest single-term denotations that are derived as much as possible from everyday language. At first sight, single-term denotations may seem unlikely to change usage in any way that helps specialists or is accessible to outsiders, but many of these terms denote what excessively complicated terms only promise to do.

Although Michael Smithson (1993, 134) observes that research on unknowns has not been extensive, he notes that there has been a rapid increase in research into uncertainty in science, which has led to an accelerating "turnover of what constitutes established scientific knowledge or truth." Another reason for a growing interest in the unknown is that scientific activities are increasingly intermingled in society (see the previous chapter), so that the institutional borders between the scientific production of knowledge and the application of scientific knowledge in the real world sometimes appear blurred. At the very least, the boundaries between science and society and between experiment and application have become ambiguous.

Among many others, authors like Ulrich Beck and Anthony Giddens have pointed to unintended consequences in contemporary society. For Giddens, knowledge is the medium of reflexive modernization. For Beck, nonknowledge is that medium, since the unintended side effects of modernization can be regarded an expression of increasing nonknowledge, or what Beck called *Nicht-Wissen* (Beck 1996).[5] In Beck's view, there are two types of nonknowledge—a nonknowledge that one does not want or need and a nonknowledge that cannot be known. For Beck, the limits of knowing are becoming ever more important in what he calls *second modernity*—a structural and epochal break that is indicated, for instance, by an increase in the significance of nonknowledge because of the rise of knowledge. Other authors, like Brian Wynne (1992), have argued that accounting for unknown dynamics and variables seems impossible and is a more fundamental obstacle to today's risk assessment than the inability to analyze known interactions accurately. Central to this stream of thought is the possibility of shifting away from traditional research strategies of reducing ignorance and toward a greater capacity to cope with ignorance (Hoffmann-Riem and Wynne 2002). This is a shift that scientists, policymakers, and the public have begun to acknowledge—that potentially harmful consequences cannot be established reliably by further research since they fall into the domain of ignorance.

As was discussed earlier, among classical sociologists, Georg Simmel shows the keenest eye for the unexpected events and surprising turns that take place in almost all fields of social life. Simmel also offered some of the most sophisticated concepts of *Nichtwissen* for the structuring of modern life, although he never connected these with his reflections on surprising effects in objective culture. This is what I now try to do. In the following, I aim to connect Simmel's reflections on unintended consequences with his writings on nonknowledge. To Simmel, trust serves as a bridge between knowledge and nonknowledge as a creative principle. Trust in Simmel's view is understood as a mechanism that glues together knowledge and the unknown.

This is the context in which the difficulties of grasping the unknown in sociology seem to have begun. In the English version of Simmel's essay on "The Secret and the Secret Society," which appears in the original German as a chapter in his *Soziologie* (Simmel 1992), the translator Albion Small sometimes translates *Nichtwissen* as *nescience* (e.g., Simmel 1906, 444, 448) and sometimes as *not knowing* (e.g., Simmel 1906, 450) without a discernible logic.[6] In the seventeenth century, nescience was sometimes regarded as what cannot be known in advance and sometimes as God's knowledge (which people cannot know and must not attempt to know). In some streams of philosophy of the nineteenth century, this led to agnosticism, a doctrine that claims that nothing is actually knowable.[7]

The symmetry in the German word *Nichtwissen*, which denotes that there can be knowledge (*Wissen*) about what is not known, is not captured in the English word *nescience*, which, according to the *Oxford English Dictionary*, today means "absence or lack of knowledge." The term *nonknowledge*, which is unusual in today's everyday English, can be traced back to the sixteenth century, whereas *nescience* dates from 1625, according to the *Oxford English Dictionary*. The etymology of *nescience* is the late Latin *nescientia*, from *nescient-, nesciens*, which is the present participle of *nescire* ("not to know"), from *ne-* ("not") and *scire* ("to know"). Thus understood, *nescience* belongs to a fundamentally different epistemic class from *nonknowledge, ignorance,* or terms such as Kerwin's *known unknowns*. No one can refer to their own current nescience because it is not part of their conscious (and thus socially constructed) nonknowledge. At most, people can refer to someone else's or their own earlier nescience. Hence, a sociological observer can ascribe nescience only in retrospect. Accordingly, I prefer the more neutral notion of *nescience*, which in its etymology and epistemology already clearly indicates its difference from terms such as *unknown ignorance* or *unknown unknowns*.[8]

Moreover, a literal translation of the word *Nichtwissen* would be *nonknowledge*, a term that was rarely used in English-speaking sociology until the 1990s, when it began to appear in articles that were written by authors whose native language was German. It can be assumed that in most cases, *nonknowledge* in English-speaking sociology was a literal translation from the German *Nichtwissen*. Overall, it appears that German authors writing in English use the term *nonknowledge* as a literal translation of the German *Nichtwissen* and that English-speaking authors use *ignorance* when referring to the opposite of *knowledge*.[9] However, most authors today have used the term *nonknowledge* to indicate a type of knowledge about the unknown. Furthermore, as is discussed below, Simmel used the word *Nichtwissen* in a sense that comes close to some English-speaking authors' understanding of *specified ignorance* as well as *nonknowledge* as used in debates on risk and the knowledge society (cf. Beck 2000; Stehr 1994, Strulick 2004).

In general, Georg Simmel saw nonknowledge as an important part of the relationship between what he called *objective culture* and *subjective culture*. However, this objective culture sometimes opposed subjective forces, which, in Simmel's writings, is the tragic conflict that permeates all domains of modern society (see chapter 2). For Simmel, the rift between objective and subjective culture can be bridged by trust in nonknowledge: "trust, as the hypothesis of future conduct, that is sure enough to become the basis of practical action, is, as a hypothesis, a mediate condition between knowledge and nonknowledge" (Simmel 1992, 393; cf. Simmel 1906, 450).[10] Trust has become an increasingly important variable in studies on risk management, decision making, and organization studies, but scholars from different disciplinary backgrounds do not agree on what *trust* actually means (cf. Siegrist, Earle, and Gutscher 2007). As authors from Luhmann to Giddens remind us, trust requires an actor or a party to enter into a position of contingency. The actor or the party involved then becomes vulnerable to the possible opportunistic behavior of the other party. Placement of trust also allows activities to be conducted based on incomplete information (Coleman 1990, Strulick 2004). Giddens, in par-

ticular, has argued that the main sources of trust in organizational procedures are "standards of expertise." Many authors also differentiate between
personal trust (sometimes referred to as *reliance*), which is an expectation
of individuals and their discretion, and *impersonal trust*, which is an expectation of institutions and the actors in them in certain situations (Giddens
1990).[11]

Simmel illustrates the importance of nonknowledge and trust with the
example of secrecy. Simmel calls the secret—which he characterizes as
becoming "effective through negative or positive means of concealment"—
as "one of the greatest accomplishments of humanity" (Simmel 1906, 462)
and notes fundamental changes in the relationship between things
unknown and known in the modern world. Many things that were formerly rendered secret (e.g., sexual intimacy) have become more public,
and experiences such as face-to-face contact with strangers (for example,
via seating arrangements in coaches and now trains) have become more
anonymous since people tend to keep their "public privacy" secret, even
in public places such as trains. He called for "a right of secrecy" (Simmel
1992, 406; cf. Simmel 1906, 462). Even more so, using the example of an
attraction to another person, Simmel points to the danger (*Gefahr*) "to
which the unrestricted possibilities of intimate relationships seduce" (ibid.).
For Simmel the lack of what he called "reciprocal discretion" can lead to
the failure of many marriages—"that is, they degenerate into a charmless,
banal habituation, into a matter of course which leaves no room for surprises" (ibid.). Generally, Simmel suggests that keeping things secret is a
structuring principle of everyday life. This is where in his conceptualization
the notion of nonknowledge (*Nichtwissen*) comes in. He writes that "in
view of our accidental and defective adaptations to our life conditions,
there is no doubt that we cherish as much truth and as much nonknowledge and attain so much error as is useful for our practical purposes"
(Simmel 1992, 385–386). To be social beings that can successfully cope
with their social environment, therefore, people need clearly defined
realms of unknowns for themselves, which Simmel called *nonknowledge*.
Nonknowledge thus is understood as a special form of the unknown that
is not lack of knowledge, error, or mere ignorance but a clearly defined
realm of what is unknown.

This is not a trivial or merely theoretical point. Consider the recent
debates on the right not to know—in Germany, as "the right to nonknowl-

edge" (*Recht auf Nichtwissen*)—about genetic vulnerabilities to diseases. In preventive genetic diagnostics, individual genetic variations are tested to identify genes that are associated with various medical conditions, since some variations can make people susceptible to some diseases and more resistant to others. At first, medical practitioners thought that they could use this knowledge to prevent illnesses in their patients. But over the last twenty years, patients have insisted on a legal right *not to know* their own genetic configuration. In preventive diagnostics, some genetic peculiarities can lead to certain illnesses, especially with increased probability at a higher age (cf. May 2003; Taupitz 1998; Wehling 2006). For many patients, the prognoses are made over very long periods, and in many cases it remains unclear if a certain disease will occur at all. For some people, the boundary between health and sickness can become blurred since a test result interpreted as negative or unfavorable can affect an otherwise healthy person's subjective well-being in everyday life. People may feel depressed or anxious about their results. When people are given the right not to know possibilities, it is not ignorance in the sense of being dumb or erroneous since people are very much aware of what they are doing. It is more a type of negative knowledge since people try to keep some of the looming knowledge out of their heads. The right to nonknowledge in personal health-related information has consequently been included in human rights documents and in national legislation since the middle of the 1990s. In the German Parliament's Inquiry Commission on Rights and Ethics, the right to nonknowledge is based on the same legal basis as the right to knowledge. In some cases, the right to nonknowledge can affect more than one individual, as when genetic testing creates tension within a family because the results can reveal information about other family members in addition to the person who is tested (Ebeling 2002). In general, recent research on prenatal testing has shown that things that are not known are more often the most important for the people involved (Jones 2007). Since the possibilities for curing many genetic diseases are very limited, preventive genetic diagnosis affects persons' current life without delivering useful perspectives on action. The right to nonknowledge thus would ensure that nobody should unwillingly gain knowledge about his or her own genes and chromosomes. This right appears to oppose the traditional idea that more knowledge is the foundation to progress. Thus understood, nonknowledge, as Peter Wehling summarized, "is not to be devalued as lack of information or a moral deficit but—perhaps for the first time in modern societies—has become an object of legal protection that should help to contain the risks and ambivalences of (scientific) knowledge" (Wehling 2006, 327).

Some one hundred years earlier, Simmel saw nonknowledge as a normal phenomenon. In his analyses of an accelerating modern society, objective culture in general was characterized by increasing nonknowledge: the "objectification of culture has sharply differentiated the amounts of knowledge and nonknowledge" (Simmel 1992, 394; cf. Simmel 1906, 450). New unintended side effects thus develop via a widening rift between knowledge and nonknowledge, which calls for more trust among individuals as they interact with each other and with the nonhuman world (such as modern technologies).[12] Like Herbert Simon's concept of "bounded rationality" (Simon 1957; cf. Gigerenzer and Selten 2001), which tries to connect knowledge and rationality, I propose a notion of nonknowledge that connects rather than opposes knowledge and the unknown.[13] This approach also looks at the unknown in a more neutral way without the often "negative excitement" surrounding discussions on ignorance.

Studying the Other Side of Knowledge

Discussing the possibilities of reaching a success in an adventure, Simmel stated that "nonknowledge about the success of a chosen path does not merely mean a quantitatively reduced certainty, but it leads to an internally and externally unique direction of our practice" (Simmel 1998, 30). The adventurer, Simmel continues, treats "the unpredictable elements of life in the same way as other people treat the predictable" (ibid.). However, modern life "transforms all of us into adventurers" (ibid., 37). In other words, capricious and unpredictable changes become the norm in modern life.

In his now classical sociological analysis of the concept of unanticipated consequences, Robert Merton (1936) tackled this theme—mainly in reference to scientific activities. Merton identified five sources of unanticipated consequences in scientific research—ignorance, error, basic values, the so-called imperious immediacy of interest, and the self-defeating prediction. He later elaborated the centrality of ignorance in that he detected two types of ignorance—unrecognized and specified ignorance. Specified ignorance is "a prelude to newly focused inquiry" (Merton 1987, 8). Implicitly, Merton thus believes in a linear development in the growth of knowledge, although he sees that new knowledge always brings an awareness of more ignorance (ibid., 8–9). He does not seem to have been interested in the fact that new knowledge also can develop into unspecified ignorance or other forms of knowledge.

Merton's "specified ignorance" can also be related to the fifteenth-century Catholic bishop and philosopher Nicholas of Cusa (Nikolaus von Kues), who spoke of "learned ignorance," which is like knowledge of the sun possessed by a person who can see. Whereas a blind person is completely ignorant, a sighted person has tried to look into the sun, but it is too bright, so he knows what he cannot know. In the sense of the blind person who is completely ignorant, the economist Neil Kay (1984, 74) also defines *ignorance* as "the absence of all knowledge." In their essay on "Some Social Functions of Ignorance," Moore and Tumin (1949, 788) (who, like Merton, stand in the functionalist tradition) define *ignorance* "as simply referring to 'not knowing,' that is, the absence of empirically valid knowledge." Furthermore, Moore and Tumin want to keep ignorance "distinct from 'error,' whether of fact or of logic, and from the act of *ignoring* what is known" (Moore and Tumin 1949, 788 n. 4; cf. Smithson 1989, 221). Moore and Tulmin thus use the term *ignorance* in a specific sense that excludes the act of not wanting to know what is already known (cf. Schneider 1962). It remains unclear whether they refer to their own knowledge or other sets of knowledge that are potentially available to an ignorant person. Moore and Tulmin's overall goal, however, is similar to Simmel's—to rescue the notion of ignorance from something negative or dysfunctional and to present it as something useful and potentially positive for the maintenance of social order.

In general, terms like *ignorance* or *nonknowledge* are used when referring to any type of unknown outcome. As Smithson (1989) does, Stocking (1998, 166) defines *ignorance* as including "absence of knowledge, probabilistic uncertainty, inaccuracy, irrelevance, and other sources of not knowing."[14] Unlike the notion of risk, where probabilities are known, and in contrast to uncertainty,[15] where probabilities are not all known, nonknowledge or ignorance can be referred to as a realm that escapes recognition (cf. Faber and Proops 1998, 128–129). Dealing with ignorance thus understood clearly differs from risk taking or risk limiting, since the risk that a certain event will occur allows both the type of possible events and their probability to be known, which thereby allows the risk to be quantified. A problem with this notion of risk is the opposition between risk taking and risk limiting. Many people are aware of risks and take many precautions, such as living in places where they think they will be safe from crime or safe from natural disasters. Moreover, not everyone has a taste for acknowledging capriciousness or, for that matter, much reason to do so. This phenomenon can be framed by an application of the precautionary principle, which states that a new process or product should not be introduced until we have convincing evidence that the risks are small and are outweighed by the benefits (cf. Saunders 2000). This argument is in accordance to the assumption that risk avoidance is the more rational strategy and risk taking is the irrational one. I believe both sides cannot and should not be assigned to such normative presuppositions.

Although many definitions of *risk* vary by specific application and situational context, the differentiation above is based on the most broadly accepted understanding of risk as the probability of a harmful event that is multiplied with the amount of expected harm that the event will inflict. Such an approach can easily lead to the questionable attitude that risk assessments deliver riskless knowledge about risk. Indeed, risk assessments involve a statistical application of theories of probability and claim to provide a rational basis for objective decision making under conditions of controlled uncertainty. Thus understood, *risk* refers to a situation when the system behavior is well known and the chances of anticipated outcomes can be quantified by probability distribution (cf. Faber et al. 1993; Knight 1921; Rosa 2003; Wynne 1992). In reaction to this notion of risk that is based on costs, which can be calculated beforehand and traded off against the advantages, some streams of sociology have recommended moving beyond probabilistic approaches to risk. This has led, for instance, to notions such as Raymond Murphy's "unperceived risk" to denote that there is risk whose "visibility often occurs after a time lag or after a disaster" (Murphy 2009, 36)—that is, risk that can be calculated after the fact as if one would have known beforehand. Although this goes contrary to classical definitions of *risk*, it outlines an important step to moving beyond probabilistic risk assessments. As a concept, it proves useful only in retrospect to outline what one should have known beforehand. It does support the saying that "hindsight is easier than foresight."

Among sociological attempts to move beyond probabilistic risk assessments, representatives of Luhmannian systems theory have been vocal critics. In Luhmann's framework, the notion of risk is inseparable from ideas of hazard, danger, and the unintended. Quite often, *risk* is defined in opposition to *hazards*, which are actively assessed in relation to future possibilities. Giddens (2000, 40) confirms that risk "comes into wide usage only in a society that is future oriented—which sees the future precisely as a territory to be conquered or colonized. Risk presumes a society that actively tries to break away from its past—the prime characteristic, indeed, of modern industrial civilization." The importance of risk is based on the dependence of society's future on decision making, which has increased and dominates ideas about the future. Niklas Luhmann notes that the problem of prevention is actually the mediator between decision making and risk taking. Among the many definitions of *risk*, Luhmann's notion sticks out because of his clear differentiation between risk and danger: "Even if it is only a question of danger in the sense of natural disaster, the omission of prevention becomes a risk" (Luhmann 1993, 31). In this way, one can conclude that today all dangers, as soon as they have been communicated as dangers, have become risks.

In general, Luhmann (1993) defines *risk* as a decision that may be regretted if the possible loss that one hopes to avoid nevertheless does occur. So far Luhmann's notion of risk coincides with definitions from authors who normally are considered not system-theory compatible. Ortwin Renn (2008, 1) has defined *risk* broadly as "the possibility that an undesirable state of reality (adverse effects) may occur as a result of natural events or human activities." However, Luhmann aims to move beyond both the traditional definition of *risk* as an opposition between risk, uncertainty, and safety, as well as general statements pointing to the contingency of all decisions. For Luhmann, risks are created permanently by conscious decision making, whereas dangers (Luhmann's counterpart to risks) are events that the affected parties have not caused themselves. Although the Luhmannian attempt to free the concept of risk from probabilistic notions appears to be comprehensible, danger as its conceptual opposite, although useful at first sight, eventually appears less than satisfactory. Besides the counterintuitive usage of the term compared to many other disciplines as well as everyday life, it is empirically difficult to attribute danger to a nondecision. With such a broad notion, both in terms of a general sociological concept and an everyday understanding, not much is gained analytically.[16] When taken out of the context of Luhmannian systems theory, what, after all, would then not be a risk-dependent decision?

In this context, Sellke and Renn (2010) discuss how *risk* always refers to a combination of two components—the likelihood or chance of potential consequences and the severity of consequences of human activities and natural events. *Ignorance* (or different types thereof), however, applies where future states cannot be clearly defined and sometimes not even be imagined. Furthermore, as Frank Wätzold (2000) has demonstrated, practical concepts (as in economic decision making), which concentrate on the reduction of risk, may indeed lead to increased ignorance. In turn, a concept may be efficient in the context of risk management but inefficient when ignorance exists. In short, despite the many definitions of *risk* and *uncertainty* (whether risk is understood as a believable probability distribution that can be assigned to possible outcomes or as the perception of future losses as the consequence of a clearly attributable decision), ignorance falls outside of the realm of risk.

The issue appears to be even more difficult with the notion of uncertainty. Funtowicz and Ravetz (1990, 87–88) define *ignorance* as the "deepest" of three sorts of uncertainty—inexactness, unreliability, and ignorance. Uncertainty thus is a certain measure of our ignorance. It is something that is not known well enough, that is undecided among stakeholders, or that is not determined. Brian Wynne (1992, 114) also talks about risk when "the system behavior is basically well known, and chances of different outcomes can be defined and quantified by structured analysis of mechanisms and probabilities." When the important system parameters are known but the probability distributions are not, Wynne talks about uncertainties. Unlike Funtowicz and Ravetz, Wynne shies away from the idea that uncertainty exists on an objective scale reaching from risk to ignorance. Instead, Wynne suggests that risk, uncertainty, and ignorance overlay each other, suggesting that ignorance can be embedded within other forms of unknowns (Wynne 1992, 116).[17] When human activities interact with the natural world (and elsewhere), however, it often appears that decisions are being made based on a clear definition of what is known and what is unknown—that is, on situations that are not necessarily communicated as lack of certainty or as a state of having limited knowledge. The crucial task is to present ignorance as a confident basis for action. These cases of decision making are of importance here, and they are common compared to dealing with uncertainty and risk in the traditional sense.

Departing from these debates, Karin Knorr Cetina adds the concept of "negative knowledge" to the collection of terms. In her view, *negative knowledge* does not mean "nonknowledge, but knowledge of the limits of knowing, of the mistakes we make in trying to know, of the things that interfere with our knowing, of what we are not interested in and do not really want to know" (Knorr Cetina 1999, 63) (and, I would add, perhaps are afraid to know).[18] In other words, some consequences of an activity might be anticipated but are judged unimportant or unlikely to be severe. Negative knowledge can also be related to what Tannert, Elvers, and Jandrig (2007, 893) have labeled the Galileo effect. This is a reference to Bertolt Brecht's play *Galileo Galilei*, in which a cardinal refuses to look through a telescope to avoid having to accept the knowledge that the planets revolve around the sun (Brecht 1980). Further steps in a research project, however, can easily force the actors involved to transform their negative knowledge into ignorance. In Knorr Cetina's approach to analyzing scientific decisions, the limits of knowing are admitted by bracketing out certain areas of knowledge and nonknowledge. On the negative side, this can lead to what David Hess has termed "undone science": knowledge could be produced based on clearly defined ignorance, but it is not pursued further (for example, because a certain strand of science industry will not invest in products that are not patentable). Hess (2007, 22) even says that there exists a "systematic nonexistence of selected fields of research." Overall, Hess situates the generation of undone science at the intersection between state, industry, and social movements. In this context, Frickel (2008) uses the term *knowledge gaps* to describe organizational outcomes of undone science. In his work on measuring the effects of Hurricane Katrina on New Orleans, Frickel uses the term in a quasi-geographical sense to refer to missing pieces in the assessment of the damage of the hurricane due to nonstudy. This can be a crucial issue since, as Hess has shown, social-movement organizations often do not have the research results available because they are unfinished, and projects are not finished because the research has not been funded. In the long run, however, the fact of "undone science" can also lead to an acknowledgment of the nonknowledge related to it, so it can suddenly be taken seriously and may even be seen as fundamental at a later point in time.

Knorr Cetina's notion of negative knowledge appears to be similar or even the same as the term *closed ignorance*, which is used by the economists Faber and Proops (1998, 117). In their understanding, *closed ignorance* means that "we either neglect problems themselves, or do not take notice of intuitive insights, experience, information, models and methods of solution which are available inside of society" (Faber and Proops 1998, 117).[19] I prefer the term *negative knowledge* since it avoids a problem that is inherent in meanings of the term *ignorance*. There is a common-sense distinction between actively ignoring and passively being ignorant. Active ignorance is a declaration of irrelevance (cf. Smithson 1990, 209), which is part of what Knorr Cetina has labeled *negative knowledge* and thus is distinguished from the two very different notions of *ignorance* as it is used in English. Since Knorr Cetina is referring to knowledge of the limits of knowing, the term fits better as a category separate from general ignorance. The term *negative knowledge* also suggests that a clear distinction between the known and the unknown may be obtainable theoretically but often is empirically difficult to define since both sides often appear in twilight, where, depending on the path of interpretation, they can be assigned to the one or the other. Everyday life is filled with many different shadings of actively ignoring things, learning to ignore, and not publicly acknowledging something (e.g., taboos or rules of irrelevance). Eviatar Zerubavel (2006, 23) explains that "ignoring something is more than simply failing to notice it. Indeed, it is quite often the result of some pressure to actively disregard it. Such pressure is usually a product of social norms of attention designed to separate what we conventionally consider 'noteworthy' from what we come to disregard as mere background 'noise.'" Because of this tendency, in the following I refer to *ignorance* as a generic term. *Nonknowledge* and *negative knowledge* will serve as specifications of knowing something about the unknown.

Unlike the approaches above, Niklas Luhmann's systems theoretical perspective uses *nonknowledge* to refer not to a lack of knowledge but rather, as Tacke (2001, 295) put it, to "a social construction, which is dependent on knowledge as its respective flip side. Experts, for instance, specify nonknowledge according to existing knowledge, causal theory, and methods." Luhmann (1992) referred to the covariance between knowledge and nonknowledge as an "increase in contingency"—that is, what is neither necessary nor impossible and what can be but does not have to be. Here, nonknowledge is regarded as the "natural" other side of knowledge—"an unmarked space"—and consequently as the other half of a distinction (cf. Japp 2000; Luhmann 1992; Willke 2002). Peter Wehling, who has delivered a comprehensive critique of Luhmann's notion of nonknowledge as an unmarked space, argues that the system-theoretical perspective is based on a static differentiation between knowledge and the unknown, which means that important questions on how to overcome scientific ignorance cannot even be asked (Wehling 2006, 196). Furthermore, as different as these ideas on nonknowledge are when compared to usages outside of systems theory, they still treat ignorance and nonknowledge—sometimes even lack of knowledge—as synonyms (e.g., Japp 2000, 225). Most often, authors simply translate Merton's terms of *specified* and *unspecified ignorance* into the German as two types of *Nichtwissen* (e.g., Böschen and Wehling 2004, 42–43; Luhmann 1992; Wehling 2006). This blurs the connotation in the original as well as the current meaning of the German *Nichtwissen*. Reminiscent of an early attempt by Weinstein and Weinstein (1978), Wehling (2006) pleads for a type of precautionary objective nonknowledge—that is, a concept of total unawareness of nonknowledge, where *nescience* might have been the more apt term.[20] The question of how the paradoxical call for knowing unknown ignorance should be handled empirically in any meaningful sense, however, remains unanswered. After all, how can you know what you do not even know that you do not know it?

Despite the problems with a system-theoretical approach to ignorance that Wehling has pointed to, it might be useful to refer to the Luhmannian notion of nonknowledge as the ordinary or normal other side of knowledge since it helps to free the notion of nonknowledge from its pathological connotation. If ignorance and nonknowledge are rendered as a normal part of life and not, as Beck (1999, 141) referred to it, as a disadvantageous "inability to know," then we need to learn to deal with it. Taking a Simmelian view, it might be useful to cope with it, since it can be understood to be an essential ingredient of progress. Knowledge and ignorance are thus to be conceptualized neither as competing perspectives in our analysis of modern knowledge societies nor as respective endpoints of an optimistic and pessimistic variant of social analysis (Stichweh 2004). If it is normal, then actors already know how to deal with it, but analysts have not yet developed the right concepts and terminology to talk about it properly.

Dynamics of the Unknown

Based on the above summaries and extrapolations of the debates on ignorance, I now suggest a preliminary categorization of notions of the unknown. First, I suggest that the English term *ignorance* can function as a kind of cover term that points to the limits of knowing, including the intentional and the unintentional bracketing out of unknowns. To grasp the latter two categories, I suggest that the term *nonknowledge* (as a literal translation of the German *Nichtwissen*) needs to be further established. *Nichtwissen* is the original term of classical sociologist Georg Simmel that refers to a type of knowledge where the limits and the borders of knowing are taken into account for future planning and action. Nonknowledge is present when an actor or a group of actors does not have sufficient knowledge about a certain question or a problem to be solved. However, the actors involved well know the point of reference of the unknown. Thus understood, nonknowledge as the acknowledgment of ignorance can be empirically identified in the analysis of documents and interview sequences when an utterance of an actor expresses the decision to act (often in a reconstructed activity chain) in light of an event that was communicated as "surprising."

To come to terms with unknowns beyond the realm of risk, I suggest a second subtype of ignorance, which I call *negative knowledge* (Knorr Cetina)—that is, an active consideration that to think further in a certain direction will be unimportant (see table 3.1). For the next category, I introduce a term that points to topics that have been referred to above—the development of new or extended knowledge, based on planning, tinkering, or acting in the face of nonknowledge. For the acceptance and continuity of research projects that are taking place in the wider society (mode 2 type of science), nonknowledge must be communicated as likely (or unlikely) to be transformable into proper new and "extended" knowledge—and under what circumstances. However, as the cases in the subsequent chapters reveal, continuity in spite of nonknowledge seems to be the goal. In that sense, successful projects are also causally dependent on a clearly

Table 3.1
Categorization of different unknowns and extended knowledge

Nescience	Lack of any knowledge: a prerequisite for a total surprise beyond any type of anticipation; can lead to ignorance and nonknowledge but belongs to a different epistemic class
Ignorance	Knowledge about the limits of knowing in a certain area: Increases with every state of new knowledge
Types of specified ignorance:	
Nonknowledge	Knowledge about what is not known but taking it into account for future planning
Negative knowledge	Knowledge about what is not known but considered unimportant or even dangerous; can lead to nonknowledge (related to undone science)
Extended knowledge	Alternatively, new knowledge: based on planning and/or research with nonknowledge; can lead to new ignorance by uncovering limits of the newly gained knowledge

stated "negligence" that allows side effects to happen. They might not be explicitly wanted, but they are known to be unavoidable. In other words, the strategy is to continue even though the risks cannot be known before implementation. Extended knowledge that develops out of nonknowledge can lead to the social awareness of, for instance, new nonknowledge by uncovering limits of the newly gained knowledge. Any new or extended knowledge can also reveal that earlier ideas on reliable and accepted knowledge must be reinterpreted. For instance, during the last thirty years, meteorologists have been able to deliver more and more detailed models for weather forecasts (cf. Fine 2007). At the same time, they have realized that these superior models open up new knowledge to perceive more unknowns (nonknowledge) for reliable weather forecasts. These two streams of knowledge generation have developed side by side without any expectation that they will merge in the near future.

The word *nescience*, which was sometimes incorrectly used as a translation of Simmel's *Nichtwissen*, should rather be seen as a prerequisite for a total surprise beyond any type of anticipation. Nescience, as a total lack of knowledge, at first sight comes close to what Kerwin (1993, 179) has termed *unknown unknowns*—things that people do not know that they do not know. It could also fill the place of Wehling's (2006) description of a complete unawareness of nonknowledge since this unawareness can only be made "visible" in sociological analysis when, like knowledge, its social utterances, constructions, or negotiations can be registered. However, as mentioned above, *nescience* belongs to a fundamentally different epistemic class from *ignorance* since nescience can be detected only in retrospect. This is why I prefer this epistemologically unique term instead of compositions like "unknown unknowns" or the rather awkward expression of a "complete unawareness of nonknowledge." In empirical studies, the term *nescience* can be used only by a godlike (social scientific) observer who already knows about the nescience of his or her object of study. More likely, *nescience* can be used as a category for reconstructing past events—for instance, in historical sociological studies where the lack of knowledge of a person or a certain group was crucial for the development of a certain technological device, as Stefan Böschen (2000) has done in his study on CFC, DDT, and Dioxin.

Nescience also can be a basis for understanding ignorance, negative knowledge, nonknowledge, and new extended knowledge. This should generally point to the dynamic character of all kinds of knowledge and nonknowledge production, a point that many of the debates on the theme have neglected or have only implicitly touched on to this date. As Brian Wynne has phrased it, any uncertainty or ignorance can be defined only "by artificially 'freezing' a surrounding context which may or may not be this in real life" (Wynne 1992, 116). In the following, I illustrate how the highlighted connotations and meanings of notions of the unknown can be linked dynamically so that they can be used as tools for analyzing different fields of the unknown.

The categorization of unknowns in table 3.1 can be seen as framed by two core types—ignorance and nescience. If knowledge is a belief that is justified as true, based on nescience a surprising event can occur that has been beyond the possibility of any expectancy and anticipation of the involved actors' knowledge. The retrospective recognition of nescience can lead to a state of ignorance—that is, a type of knowledge about the limits of knowing. This is where an observer can register the social understanding of an unknown phenomenon. Actors then can decide to develop this ignorance to what Merton has coined "specified ignorance." However, the empirical reality often shows that specified ignorance can have at least two diametrically different meanings based on the reaction and evaluation of an observed event—nonknowledge and negative knowledge. At its most basic, the term *extended knowledge* means knowledge at a certain time $t +$ 1, which can also mean new knowledge, which is able to substitute older sets of knowledge.[21] However, extended knowledge can also inherit new knowledge about further gaps in knowing and thus can function as the precursor of learning about new ignorance and nonknowledge, as is depicted in figure 3.1.[22] Indeed, for reasons of clarity, figure 3.1 is dynamically simplified. For instance, unknowns can increase when people act based on what they think is reliable knowledge (cf. Dovers, Norton, and Handmer 1996, 1148; Wynne 1992, 123), but this knowledge will turn out to be wrong. I do not conceptually pursue this aspect since it is almost impossible to detect this type of "unknown" empirically (mostly it is part of what I have coined *nescience* and can be realized only in retrospect). In this context, it also needs to be noted that variables like nonknowledge or ignorance are autocorrelated over time because what people know and do not know always depend on what they knew and did not know earlier.

Figure 3.1 illustrates how these different types of unknowns are linked together dynamically so that they do not appear as artificially frozen and as objective givens. The figure does not suggest that knowledge production or evaluation always begins with nescience, even though nescience is positioned on the top of the figure. In many areas of social life, planners anticipate the borders of planning and of the possibilities for a realization of a certain plan. The arrows in figure 3.1 are conceptual linkages that *can* nevertheless have causal connection. The arrows from ignorance to its two subtypes, for instance, are to be understood as special cases of ignorance that are preceded by ignorance. However, negative knowledge and nonknowledge can also causally develop out of ignorance. The dashed arrow near further nonknowledge should indicate that the search for new knowledge could potentially lead to more and more unknowns instead of new knowledge because formerly valid knowledge can be found to be invalid. Hence, ignorance, nonknowledge, and negative knowledge can all stand at the defined beginning of an activity as well as on the beginning of a sociological observation. The model also indicates steps that can be undertaken or ways that link together different types of unknowns, but it does not decide on concepts and practices to accomplish the stages in development of knowns and unknowns. Rather, the different stages can be used to identify the complexity—that is, an intricate combination of parts com-

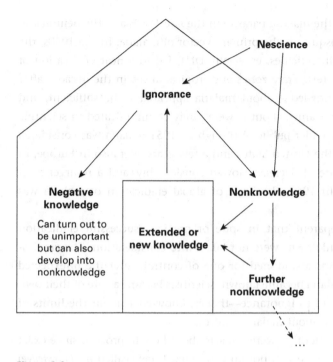

Figure 3.1
The house of the unknown

pared with earlier phases in a development process (cf. Nowotny 2005; Law and Mol 2002)—of the development to new or extended knowledge as well as new forms of the unknown.

To account for the terminology introduced with "the house of the unknown" in figure 3.1 and to demonstrate that a manageable model of how these unknowns work can be derived from it, in the next section, I use the example of malaria control to show how these different types of unknowns can develop in research on widespread vector-borne disease. This contributes another important element to an approach for dealing with and analyzing in depth the cases of ignorance and surprise that are presented in the following chapters.

More Questions Than Answers: The Case of Malaria Control

Malaria, one of the world's most important vector-borne diseases, is caused by several species of parasites. Systematic control of malaria started after the discovery of the malaria parasite in the late 1880s and the demonstration that the mosquito is the primary vector of malaria. In the 1950s, the use of different insecticides, especially DDT, led to a near eradication of malaria in many temperate zones and tropical areas. In the decades after World War II, knowledge about malaria appeared to be sufficient, and questions of uncertainty about it were rarely communicated in scientific journals or the broader public. Although by 1951 malaria was considered eradicated from the United States and a few years later also in Europe, in the 1960s the rate of decrease slowed considerably, and a resurgence of cases occurred. In 1969, the hope of global eradication of malaria was finally abandoned.

It became apparent that in spite of previous successful eradication efforts, crucial unknowns were not taken into consideration. The eradication campaign was abandoned for one of control. After the unexpected resurgence of malaria became known, scientists became aware of their own nescience that lead to ignorance—that is, knowledge about the limits of former knowledge about malaria control.

At this point, scientists were unclear about how to proceed since exact knowledge about what was not known (what I have called *nonknowledge*) was not available. It took a few years before this ignorance was specified (Prescott, Harley, and Klein 2005, 824–826) so that it could be qualified as nonknowledge and used for future planning. The knowledge about the unknown that was generated showed that the malaria parasites had become resistant to antimalarial drugs. Although the old knowledge about eradication of malaria was superseded and new knowledge how to handle antimalarial drugs in the future more sustainingly was not yet produced, there was still a clear knowledge about what was unknown. Based on this nonknowledge, new research on how to produce "extended knowledge" could be started. The question was how to handle the fact that malaria mosquitoes become resistant against a drug.

Before that question was answered, another stream of unknown became apparent. When a person survives a malaria infection without taking an antimalaria drug, he or she develops some form of immunity to malaria. However, when insecticides reduce the number of people who are exposed to malaria, fewer people build up this natural immunity, and more people become susceptible. This knowledge about yet more areas of unknowns can be called a broadening of the range of nonknowledge. It is also based on a set of negative knowledge since for a long time the role of natural resistance of people to malaria was rendered relatively unimportant for the general dynamics of fighting malaria. Based on careful planning with these sets of nonknowledge in the 1990s (Janssen and Martens 1997), a management system was developed that took into account the fast dynamics of malarial and mosquito populations as well as the slower dynamics of malaria susceptibility in humans. This form of extended knowledge about how to control malaria can be yet another step to uncovering even more limits of the newly gained knowledge, so it will potentially lead to new nonknowledge and yet more extended knowledge. The next step was taken in 2002, when the discovery of an antibody that protects against the disease was announced (Ito et al. 2002). It was hoped, as Kwiatkowski (2005) suggests, that the genome sequence of the most deadly agent of malaria, *Plasmodium falciparum*, will provide targets for new drugs or vaccines. A new vaccine in trials was designed to cue the immune system to recognize a protein found on the parasite when it is initially injected into the bloodstream by a carrier mosquito. The vaccine was expected to enter the final stages of clinical trials in 2007 (cf. Callaway 2007). However, a short time thereafter scientists questioned the promises of the new vaccine, since treatment strategies would be continuously hampered by ever-emerging parasite resistance. Indeed, Langhorne, Ndungu, Sponaas, and Marsh (2008) claim that surprisingly little is known about this issue and that more questions have arisen than answers can be given. In other words, in the development of malaria control, ever new ignorance has resurfaced that has been and in the future needs to be specified into nonknowledge to stay tuned to the control of malaria.

In the summer 2008, ignorance was transformed into several sets of nonknowledge—such as the unknown mechanisms that regulate immune pathology in infected but partly immune humans. Langhorne et al. (2008, 730) sum up the most recent developments in malaria research as follows: "Although there have been some considerable advances in understanding the host response to plasmodium, it is not yet known what to measure as a correlate for immunity, what mechanisms regulate immune pathology in semi-immune people, what (if any) defects contribute to the relatively ineffective immunity in children and why immunity to *P. falciparum* infection can apparently be short-lived." In other words, it is already clearly known what is not known and in what direction planning and research needs to move on to, so this most advanced-candidate vaccine for malaria is ready for its biggest test. However, several clearly defined unknowns already now have been specified into nonknowledge. As Martin Enserink (2008, 1623) has reported from a meeting of the American Society of Tropical Medicine and Hygiene, even if the test is successful, it is not known whether "rather than preventing serious malaria, the vaccine may simply be delaying it by a number of years." Even before the test has been finalized, further possibilities for specifying and prospectively coping with the unknown are being fathomed.

Other cases that are discussed in more detail below call for strategies that have the potential to handle different types of unknowns based on unexpected "natural" changes. To illustrate the overall relevance of nonknowledge for decision making in research, thus far the case of malaria control has stayed on an abstract level. Once we move to the local problem areas, however, we see how successfully dealing with the unknown can be embedded in a societal learning process in such a way that surprising occurrences can be handled and used for further planning. A design for coping with different types of unknowns must be able to accommodate revisions to issues that expose new ignorance, although they were agreed on as valid knowledge in previous planning steps, as was also the case with the belief of eradication of malaria based on different insecticides. New extended knowledge can be fed into the next step to (possibly) uncover further ignorance and can potentially question the new gained knowledge yet again. Because real-world implementations like large-scale ecological design projects or the battle against malaria cannot be based on the institutional conditions of unrestricted scientific knowledge production, they have to deal with what is known and what is unknown. A clear terminology is a necessary first step to handling this task. It helps the recurrent exposure to new types of unknowns to be seen as an opportunity by acknowledging that surprising effects are perhaps unavoidable, since they fall into the domain of ignorance, nonknowledge, and negative knowledge.

The model of dynamically connected types of unknowns (figure 3.1) defines several types of knowledge beyond the immediately known. By helping to ascertain how these types could be connected and dynamically linked, the model can help avoid having one usage of a concept exclude another one or overlap in a way that blurs meanings. Instead, each unknown can potentially be the cause of another one, and each can play an important role in research and development processes. This dynamic development process shows that different types of unknowns are embedded within other types of unknowns and potentially are extensions of other types. I assume that with only minor changes many more existing usages and subcategories of knowledge about the unknown can be situated within the model (cf. Bernstein 2009; Groves 2009; Hess 2009, Kuhlicke and Kruse 2009). The typology allows other meanings, shadings, and further limbs to be added to the model. Only a few core types of unknowns have been presented here.

In the next section, I discuss how the process of piling up of nonknowledge and ignorance can be embedded within a broader conceptual foundation by discussing experiments in the knowledge society as well as terminology and conceptual parcels on surprise. The framework developed serves as a model for further thinking on the possibilities of unleashing and controlling surprising events via an "experimental" avenue of handling ignorance during the course of development. I discuss several examples of experimental strategies for coping with surprising events in science, social planning, and political decision making to improve our understanding of science in its social context, its future limitations and challenges to knowledge production, and its relation to ignorance.

Toward the Experimental Integration of Ignorance and Surprise

Whereas a faith in total control and full knowledge of ecological systems and social processes implies an ability to act only when everything is known in advance, an experimental approach allows us to accommodate different factors in spite of gaps in knowledge. Connecting the discussion back to the classical notion of society as experimentation helps facilitate the building of a community of inquirers who deliberate about local social challenges, form hypotheses about the appropriate means and ends of practice, and test their assumptions. An experimental approach, as outlined here, is thus a means to launching an environmental project in spite of unknowns. As Ludwik Fleck (1979, 86) puts it:

If a research experiment were well defined, it would be altogether unnecessary to perform it. For the experimental arrangements to be well defined, the outcome must be known in advance; otherwise, the procedure cannot be limited and purposeful.

The more unknowns there are and the newer a field of research is, the less well defined are the experiments.

Even the least defined experiments in society need to have a few defining conditions, however, to avoid becoming strategies for trial and error or for successfully muddling through. Using the points discussed so far, I propose an experimental model whose starting point is a special observation by certified experts (as in the malaria case) or by almost anybody who is member of a society (as in Robert Parks's idea of society as experiment). An observation is regarded as normal as long as the observed process follows the expectations of the observing actors. However, the complex links between natural and social factors and the rift between subjective and objective culture (Simmel) often prevent the expected path from being taken.[23] In the strategy outlined here, surprises occur when people at a certain time in a certain place communicate an event as surprising and then adapt their behavior in the future, regardless of a later communication that reveals an error or nescience. The reason for the event leading to a surprise is redefined as a mundane error in the human cognition. If such a registered surprise is taken seriously (cf. figure 3.2), then it forces the actors to question their previous knowledge of the human-and-nature system and to adjust their theories and their future actions accordingly. Nevertheless, the acknowledgment of what I have termed *nonknowledge* must proceed one step further. Otherwise, it will move into a trial-and-error process.

If a surprise leads to a complete failure of previous arrangements or even the appearance of unintended side effects, then a process of negotiation may result in the questioning of existing theories and assumptions about strategies and plans of action, which can lead to an accommodation of the original strategies. Accommodating surprising events means to adapt to changing conditions—for example, as in *adaptive management* (e.g., Holling 1986; Walters 1986). This implies the central tenet of "a continual learning process that cannot conveniently be separated into functions like 'research' and 'ongoing regulatory activities,' and probably never converges to a state of blissful equilibrium involving full knowledge and optimum productivity" (Walters 1986, 8).[24] Adaptive management became an important concept in resource management after Kai Lee (1993) used the metaphors of compass and gyroscope to connect scientific analysis and civic participation. The compass and the gyroscope, Lee (1993, 5–6) stated, are "naviga-

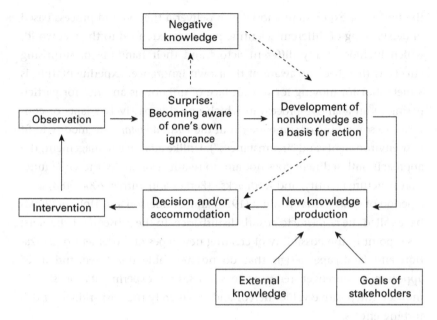

Figure 3.2
Surprise and acknowledgment of nonknowledge as prerequisites for successful interventions and learning. The solid arrows illustrate the cycle of dealing with ignorance and surprise—that is, processes where new knowledge is generated considering academic (external) knowledge as well as the stakeholders' interests and goals. The dashed arrows symbolize alternate paths.

tional aids in the quest for sustainability." They integrate science and democracy. Science links to human purpose and thus serves as a compass. Democracy is "a gyroscope: a way to maintain our bearing through turbulent seas" (ibid., 6). The scientific method (the compass) can warn when the direction is off course, while the democratic process (the gyroscope) lends stability when humans encounter turbulence in their relations with nature. Adaptive management for Lee thus is a compass in searching for a sustainable future.

Although adaptive management has been defined in various ways since its development in the 1970s, it generally is referred to as a systematic process for improving management strategies by learning from the outcomes of operational programs (Marmorek 2004). In the concept of experiment discussed here, I am not solely interested in adaptive management's application of data based on repeated measurement (monitoring). Instead, the notion of experiment should also deliver a theory- and process-based understanding of different scientific processes extended to the real world, which includes many different actors and their handling of surprising effects as they become aware of their own ignorance. Experimentation is a metaphor for moving forward in face of unknowns and not for participating in an adaptive process in which one side passively adapts or conforms to something or someone (from the Latin *adaptare*, "to modify to fit new environments and circumstances"). Unlike adaptive management, the approach outlined here does not aim to retain "essentially the same function structure identity, and feedbacks" (Berkes and Turner 2006, 488) after a period of crisis—what Holling (1978) and Gunderson and Holling (2002) have called the *resilience* to absorb disturbance. In the subsequent chapters, cases point to the possibility of creating new types of settlement organization and landscape design that do not resemble the "predisturbance" appearance. Moreover, referring to surprises via experimentation should indicate the positive value thereof and not merely the crisis-ridden or disturbing effects.

Although many authors in the field of adaptive management have contributed to our understanding of dealing with uncertainty, the general approach is still to use many small adjustments via learning by doing, reacting, and adapting as a project moves along. Understood in this way, adaptive management is an incrementally constructed approach that supports processes of development in which many small steps over time create change. This approach, sometimes referred to as *muddling through* (cf. Lindblom 1959), has been developed further in frameworks such as *intelligent trial and error* (Woodhouse and Collingridge 1993; Woodhouse 2007). Edward Woodhouse has studied incrementalism toward the goal of learning from experience and eliminating the notion of muddling through via a refutation of the lack of goal orientation, which many critics of Lindblom's and his scholars' concept have objected to in the past.[25] In a way, all processes of learning require some trial and error that leads to new discovery, but they do not appear to be very intelligent. Indeed, the notion of "trial and error" for many appears to be the wisdom of those who do not know the direction of where they need to go. The addition of the adjective *intelligent* thus does not appear to be very attractive either.

In sum, although the concepts of adaptive management and of intelligent trial and error have been inspiring for me, I believe that a notion of public experiment with a focus on the core of experimentation—fostering and controlling surprises in a modularized manner—can move us one important step further. Experimentation does not merely rest on the idea that small steps should change the status quo but attempts to account for what lies outside the sphere of risk assessments and probabilistic design— ignorance. Public experimentation, as discussed above, can be used as a label for processes for finding out a strategy that is accepted by many of the actors involved but moves away from the idea that the postsurprise order will or needs to have the same function and structure. Indeed, an experiment expedites and assists a surprise by acknowledging and (ideally) documenting what has not been known before so that the stakes are more transparent for the actors involved. Public experimentation, as outlined here, can include the expectation of both continuity and discontinuity. Both cases point to changes—those that are based on continuity via a return to early paths of development after a surprise and those that are based on discontinuity—that is, a possible setback or even failure following a surprise leading to a fundamental change (e.g., replacement) of institutional frameworks. Besides the examples in ecological design, many experiments in alternative housing projects, sustainable food consumption, or complementary currencies (cf. Seyfang 2009) explicitly have the goal to build new institutions, to reshape socio-technical infrastructures, and to create new systems alternatives.

In an experimenting society, knowledge production is supported and initiated via research in settings outside the actual research and implementation site and via knowledge from external sources (see figure 3.2). This is crucial since it involves the production of new knowledge by modifying underlying norms and objectives, which involves a major restructuring of cognitive processes and theories. Such a learning process may result in the questioning of existing theories of and assumptions about strategies and plans of action. The process can be regarded as complete when reciprocal interaction between observation and intervention closes a cycle (figure 3.2). On the one side, new knowledge enters into new arrangements, and on the other, new arrangements are fed back to produce new knowledge. This process is a prerequisite to using surprises as opportunities for improving a design process.

In part II, I use the framework developed so far (cf. figure 3.2) in more detail and discuss strategies for reaching socially responsible decisions and interventions beyond predictions, forecasts, or risk analysis. Because successful ecological implementations require socially and scientifically robust strategies and applications, I suggest that restoration strategies can be placed along two dimensions—(1) openness to surprises and (2) knowledge about the unknown or acknowledgment of ignorance.

II   Practice

4   Ecological Restoration and Experimental Learning

It is the deliberate inclusion of unwanted, "negative" elements, such as dangerous species, or elements such as fire that distinguish restoration of whole ecosystems from other forms of restorative land management.
—William R. Jordan III (2006, 26)

The call for more public access to ecology as a scientific discipline can be traced back to Ernst Haeckel at the beginning of the twentieth century and later to Aldo Leopold (Gross 2007a). However, only recently have attempts to develop a *public ecology* (Robertson and Hull 2003; Luke 2005; Ingram 2009) been successful, perhaps because they have been made not by academics or university ecologists but by various people who are involved in debates over political and ecological issues. A more exact characterization of public ecology is most often ascribed to the field of ecological restoration. This is not a coincidence. Ecological restoration and the renaturing of landscapes are inherently human activities that are based on individual and group attitudes, beliefs, and values. As is discussed above, a distinctive feature of restoration is that the human influence on the landscape and nature is not always perceived as bad. Public ecology, as David Robertson (2002) has pointed out, should help citizens, stakeholder groups, and policymakers build ecological solutions that take advantage of local knowledge and experience of what works and what does not in complex and uncertain environmental issues. By doing so, the idea of a public ecology acknowledges that scientific knowledge for real-world problems can never be perfect or complete.

In the following, I argue that ecological restoration projects can be arranged so that the actors involved can react to the surprising effects of inevitably imperfect human interventions in the natural ecosystem.

Building on the discussion in chapter 2, the relations between scientists, local citizens, and other stakeholders in processes of restoration can be modularized via recursive interdependences between the previously discussed two modes of knowledge production—mode 1, discipline-based research, and mode 2, which is exploratory and experimental research in public.

With this discussion as a reference point, I use the development of the Chicago shoreline and in particular the design of Montrose Point on Chicago's north shore to discuss how connections between the two modes of knowledge production can help people cope with ignorance and unexpected events. This case should sharpen our understanding of ecological field practice and an openness to the surprises uncovered by nonexpert participants such as birdwatchers, hikers, picnickers, and anglers.

Ecology in Society: The Shifting Boundaries of Ecological Restoration

Ecological restoration has a long history in the Chicago area. In 1977, a group of volunteers started the first amateur-only restoration project in the United States. Focused on the area near the North Branch of the Chicago River, the North Branch Restoration Project evolved into a volunteer stewardship network that today has some 8,000 volunteers. Because ecosystems in restoration projects are constantly changing, the question of what to restore or to renature becomes a moral and aesthetic as much as a scientific challenge.

Restoration and renaturing of ecosystems are normally understood as a step beyond the one-sided conservationist and preservationist strategies of traditional environmentalists who seek to protect nature. Although restoration practices can complement nature conservation efforts, restoration activities (despite the etymology of the term *restoration* and unlike many conservation programs) do not strictly reproduce historical biophysical conditions.[1] Recent debates about global climate change have supported restoration activities as being essential to providing new spaces for migrating habitats and their associated flora and fauna (Harris, Hobbs, Higgs, and Aronson 2006). Because research on social adaptation to climate change questions the static notion of nature on which many social sciences have built their implicit foundations, it appears to be challenging the wisdom of many types of ecological design.[2] In general, compared to more conservationist or protective approaches to nature, restoration is regarded as a development away from an ideal "hands-off strategy" to an active attempt to re-create, invent, design, or renature ecosystems (cf. Baldwin, De Luce, and Pletsch 1994; Gross 2003a; Friederici 2006; Higgs 2003). To interact with the landscape in activities such as ecological restoration, human actors need to accept failures and other surprising events since they are welcome sources of significant information. This approach calls for societal and scientific openness to surprising events that originate from the social or ecological system so that they are viewed as potential sources for learning. As has been shown in the preceding chapters, the consequences of these surprises cannot be reliably established by further research since they fall into the domain of ignorance. Understood in this way, successful ecological projects deliberately handle these surprises by considering the unknown—even if these surprises can lead to disorder.

As discussed in chapter 2, restoration is a good example of an in-situ and in-context mode of knowledge production. Since ecological restoration is a field of practice that also is scientific, it is particularly receptive to an experimental approach. The development of restoration ecology suggests a prolific interaction between research-oriented and application-oriented forms of science, with shifting boundaries of authority (cf. Gieryn 1999) rather than a radical replacement of discipline-based research. The reconstruction of this process tells us something about the coping with surprising events of science and its applications.

Unlike the mode 2 theory of knowledge production, which foresees a new form of science for the future, the work of Thomas Gieryn (1999) is concerned with the creation and maintenance of boundaries between science and nonscience and between disciplines that have different knowledge bases, methods, and practices. In negotiating the boundaries of science, people argue over and ultimately decide what is scientific and who is a scientist amid contests for credibility, prestige, power, and material resources. Gieryn understands *boundary work* as those moments when the question of "What is real science?" is explicitly discussed. To this end, boundary work is understood as a process of defining a social boundary that distinguishes science from nonscience and that is driven by the need to establish the legitimacy and epistemic authority of science—that is, "the legitimate power to define, describe, and explain bounded domains of reality" (Gieryn 1999, 1).

Gieryn's notion of boundary work is useful in examining ecological restoration. I'use the concept of boundary work to demonstrate that a mode 1 and mode 2 framework does not appear to be a set of "rules for proper fact-construction, but . . . rhetorical tools deployed in the pursuit or defense of epistemic authority, or in efforts to deny legitimacy to rival claims" (Gieryn 1999, 362). I also show that, since the 1970s, mode 2 knowledge production has contributed to the major developments in the field of ecological restoration. By relating the field of ecological restoration to this debate, I extrapolate over the last forty years of restoration work (rather than pick one project at a certain time) to suggest that ecological restoration has evolved in a process of circular connection between mode 1 and mode 2 knowledge production. In this process, the perimeters dividing science from nonscience are continuously renegotiated with the goal of gaining epistemic authority. Restoration is thus seen as a procedure where academic research and hands-on practice cope with ignorance and surprise, feed off each other, and are refined in the process.

The first journal in the field, *Restoration and Management Notes*, was founded in 1982 by Bill Jordan to publish articles and notes from whatever background, as long as they dealt with subjects related to restoration. It provided a forum for all kinds of articles related to restoration, including philosophical and aesthetic ones. One of the concerns of the founding editor was to have a literate style that was accessible to as many people as possible.

In 1987, the first collection of writings on ecological restoration, entitled *Restoration Ecology: A Synthetic Approach to Ecological Research*, was published (Jordan, Gilpin, and Aber 1987). The book contained mainly reflections on the novelty and use of restoration but also academic and formal science articles as well as practitioners' and artists' essays. Later in 1987, the Society for Ecological Restoration (SER) was established. The general aim of SER was to fulfill Aldo Leopold's aspiration for humans to take their place as full members of the land community (Leopold 1991). For Bill Jordan, restoration "is work and it can also be play, a way of communicating with other species and with the landscape, a mode of discovery and a means of self-transformation—a way of both discovering the natural landscape and discovering ourselves in that landscape" (Jordan 2003, 78). This also means that ecological restoration is never static. Instead, "The process of restoration itself is dynamic, adding to the dynamics of the historic ecosystem the new dynamics of its interaction with the restorationist and the human society that he or she represents" (ibid., 22). More important, "the product of restoration is an *artificial natural* community. And when you use those two words together, they cancel each other out and you begin to realize that they mean nothing when used in opposition to each other. We do this in a very literal way when we set out to 'restore' a 'natural' system" (Jordan 1990, 78).

In 1993, the world of ecological restoration began to change quite rapidly. In that year, a new journal, *Restoration Ecology*, was founded to be an academic and peer-reviewed journal that was disciplinarily bound to the science of ecology. Then in 1999, the first journal in the field, *Restoration and Management Notes*, was renamed *Ecological Restoration*. With the new journal, *Restoration Ecology*, however, the distinction between restoration ecology and ecological restoration became institutionalized. The goal was to make restoration a real science. Restoration split into restoration ecology (the science) and ecological restoration (the social and implementational practices) (cf. Higgs 2005). Thus understood, scientific reliability and social acceptability are rendered as two distinct realms. Until 2006, the instructions to contributors to *Restoration Ecology* stated that "the primary emphasis of the Journal is on ecological and biological restoration, and it also publishes papers on soils, water, air, and hydrologic functions." Although this did not explicitly exclude pieces on philosophical, aesthetic, or social issues, these aspects of ecology were not mentioned. After two years in print, no such articles had appeared except for a few attempts in the very first issues. Today, however, the guidelines for authors are much broader. The notes to authors now state that "Contributions may span the natural sciences, including ecological and biological aspects, as well as restoration of soil, air, and water when set in an ecological context; and the social sciences, including cultural, philosophical, political, educational, economic, and historical aspects."

A watershed in "boundary work" (Gieryn 1999) between mode 1 and mode 2 ecological research is the debate between Eric Higgs (1994) and Anthony Bradshaw (1993, 1994) that appeared in the then new journal. In the second issue of the journal, Bradshaw (1993) claimed that for the "hard science" faction, restoration has to be a science and a successful restorationist has to be a good scientist (Bradshaw 1993, 73). A successful scientist, Bradshaw believes, must "establish general principles," has to "carry out proper experiments to test ideas," needs to involve "careful observation," and must have "proper ecological understanding and training" (ibid., 72). Eric Higgs challenged the attempt and called Bradshaw and others' perspective a naive idea of an "austere and disengaged science" (Higgs 1994, 138) and "a narrow view of science" (ibid., 145) that could never work in the practice of restoration ecology. Instead, Higgs claimed that "restoration ecology ought to be on the forefront of an inclusive, integrated, and holistic ecosystem science" (ibid.). Higgs questioned Bradshaw's idea of "a traditional view of science that makes sharp distinctions between what is and what is not science, and [which] places scientific knowledge above other forms of understanding" (ibid., 142). He also criticized Bradshaw's failure to include any discussion of the goals of restoration in his essay: "The brilliance of ecological restoration thus far has been a fusion of practical and theoretical knowledge and a convivial and unique mingling of amateurs and professionals within the larger environmental movement" (ibid., 145). Although Bradshaw (1994) in his reply to Higgs remains guarded and claims to agree with Higgs, his understanding differs from Higgs's. Bradshaw notes that he wants ecology to be understood as science first. He nevertheless points out that ecological restoration "has to be taken into the real world where it is much needed, where it will have to work with other disciplines such as economics, politics, sociology, anthropology, and the real matters of people and their hopes and fears" (Bradshaw 1994, 147f.).

For about a decade after this debate, *Restoration Ecology* restricted itself to articles dealing with basic research and technical problems in ecosystems. The problem was apparently that the experimental attempts of the early practitioners did not find their way into peer-reviewed journals and did not allow comparative studies that were useful for ecological sites in different places. Thus, some of the ideas of the founding members were bracketed out, and some practitioners who were affiliated with academic settings left the nonscientific field. The goal was to build up an academic discipline, appropriate the catchy title of *restoration*, and leave the site-specific approaches of the practitioners. To develop, restoration needed an academic foundation that focused on basic research and was less open to surprising events stemming from field research, which included nonscientific elements.

Today, the Society for Ecological Restoration and several journals have established themselves in many universities around the world and can finally regard restoration ecology as a "real science." They have profited

from the work of the early practitioners and have better defined and
refined their mission over the years. By attributing selected characteristics
to the institution of science, they have gained greater authority than that
commanded by the original streams of restoration ecology. The old boundary that was set up by the practitioners—who wished to include aesthetics,
community well-being, and a playful "trying-things-out strategy"—was
erased by the academics, who regarded basic research in ecology as the
only real science. They thus constructed a new boundary.

Although disciplinary research in ecological restoration needs to be
implemented in the real world, the development is based on a reciprocal
or recursive process between academic and lay strategies of ecological
design and restoration, with shifting boundaries between what is regarded
as science and what is regarded as nonscience. This also implies a connection between scientific knowledge production and other types of knowledge sets. After a dearth of connection between original practice-oriented
restorations, twenty-first-century restoration practitioners and scientists
have become increasingly viewed as participating in a mode 2 type of
science, which should make progress in their collaboration with the wider
public. Increasingly, natural scientists themselves have called for a shift
away from purely inner-scientific activities (cf. Swart and van Andel 2008;
Van der Windt and Swart 2008; Weinstein 2008). Mark Davis and Lawrence
Slobodkin even argue that by continuing to frame "its goals and objectives
in a scientific context, the field [of restoration ecology] paradoxically may
actually be undermining its credibility" (2004, 2). Academic scientists have
reacted to the size and repercussions of large ecological experiments in
restoration projects, which cannot be understood and reliably accomplished without the inclusion of so-called nonscientific elements like lay
knowledge or aesthetic preferences of citizens. Stefan Halle (2007) points
out that the solution would have to be a shift in restoration sociology
toward more science, which includes the social sciences. This, as Temperton (2007) has suggested, would be a move to a transdisciplinary type of
knowledge production (see chapter 2 above) that cooperates with groups
such as social scientists or landscape artists. Indeed, the early proponents
of restoration claim that this has been what restoration was from the
very beginning (cf. Gross 2003a). In short, from a grassroots, bottom-up
movement, a surprise-averse academic discipline has emerged that in
the first decade of the twenty-first century has been seeking to reconnect

(at least a little) with its nondisciplinary origins. It appears that the non-scientific—and more surprises in public experimental settings—can be allowed back in after all.

In sum, after a phase of *scientization* of ecological restoration, a new call for the inclusion of what the other camp wanted in the first place was heard. Once the scientizers had defined the terrain as theirs, they opened small doorways to let other groups participate. A similar pattern has occurred in public health research (cf. Minkler 2004).[3] By the late 1990s, those who called for top-down, scientized work in the 1960s and 1970s and kept community-based health practitioners out of the major journals and jobs realized that their projects were not working well. Today, there is a revival of community-based public health, but not surprisingly, the add-communities-later method is proving to be much less effective than public health work based on the earlier community-involvement-from-the-beginning approach. This indicates that one of the ways that boundary work operates is to allow one group to define terrain. This group then often offers earlier contenders a (small) seat at the table, but—more often than not—without as much real power.

If we understand mode 2 as a moral program for new types of science and not as an analysis of actual changes, as some commentators have argued (see above), there is one point where the advocates of a new knowledge production have proven to be right. In the end, academic and disciplinary research cannot make progress without including the boundary negotiations with wider society and thus adding a crucial element of uncertainty to both the epistemic acceptability of scientific results and the social context that is being exposed to scientific surprises. This means that any decision about the "right" science is conditioned by the context of application and evolves with it. Put differently, in the 1970s and 1980s, ecological restoration started out as a genuinely new form of knowledge production that knowingly tinkered with ignorance and surprise, but—perhaps even because of that—it soon became the site of discipline-based knowledge production in the style of mode 1 to control the surprising results reaped by the practitioners to channel them into academic outlets. The numbers of practitioners grew but not at nearly so rapid a rate as the mushrooming field of academic restoration (cf. Young 2000). This has to do with the fact that the practitioners were focused on the long-term recovery of ecosystems, which provided useful insights into problems of academic ecology today. A form of new knowledge production that hardly any academic took seriously in the 1970s and 1980s had developed into a traditional mode of science—to the extent that it exhibits more features of traditional academic science than the purportedly transdisciplinary forms of postnormal science today.

An example from restoration and landscape design—the development of the Chicago shoreline from the late nineteenth century to the present— illustrates the connection between knowledge production and application and its dealings with ignorance and surprise. In the next section, this development with a special focus on Montrose Point, a human-made peninsula on Chicago's North Side, is discussed.[4] Montrose Point is always changing in unpredictable ways, in part because of human interventions but also because of natural changes along the lake.

New Land: Shaping the Chicago Shoreline

In 1992, the Department of Environment of the City of Chicago was established under Mayor Richard M. Daley. Since then, 1,000 acres of brownfield sites have been remediated and restored. One of the department's goals was to restore Lincoln Park on the north side of the city. The 1,200-acre (5 square kilometer) Lincoln Park stretches almost all along the shoreline of the city of Chicago. Another goal was to restore the Calumet area on the southeast side of Chicago. The Calumet area, one of the largest wetland areas in North America, was a highly industrialized region that was replete with brownfields and remnant habitats (cf. Westphal 2004). The Calumet region covers 160 square miles (414 square kilometers) of northeast Illinois and northwest Indiana but includes only 10 percent of the city of Chicago.

In many respects, Lincoln Park has been an important location for Chicago and its history. As late as the 1820s, the area now known as Lincoln Park was virtually untouched by Europeans and remained primarily forest with stretches of grassland and occasional quicksand. During an 1837 cholera epidemic, the city of Chicago took control of the land that now stretches along the park for a cemetery. In 1858, the graves were moved northward to an unused area that was named Cemetery Park. During the Civil War, portions of the cemetery were used as a Confederate burial ground, and in 1865, the graves were moved northward again. After the 1865 assassination of the sixteenth president of the United States, Abraham Lincoln, the park was renamed Lincoln Park. Today Chicago's lakefront is a well-known leisure spot, especially in the summer, and Lincoln Park provides a living outdoor laboratory with habitats that are rich in native plants, migratory birds, and aquatic life.

The largest land accretion project in Lincoln Park's history was the acquisition of the area that today is surrounded by Montrose Point some three miles (4 kilometers) north of the historical center of downtown Chicago ("the loop"). Like many other areas of Lincoln Park, this acreage was created from breakwater development and the reclamation of submerged lands from Lake Michigan, but for the most part it was built on landfill and human waste. In the early twentieth century, the area was known as Montrose Extension. In the 1990s, the general goals for the design of green spaces in Chicago were to stop the continued loss of critical habitats, to restore natural communities, and to ensure urban conservation for future generations.

In 1994, a consortium of more than thirty public and private organizations formed an umbrella organization that today is known as the Chicago Wilderness Coalition. Its goal is to restore, design, and manage the ecosystems of the Chicago region for the benefit of Chicago's citizens. Today the Chicago Wilderness Coalition has grown to over 160 member organizations.[5]

Chicago's shoreline has been extensively modified since the city's incorporation in 1833—first, for trade and transport and later, for recreation and beauty. The once nearly straight twenty-seven mile (43 kilometer) stretch of mostly sandy beaches has been transformed into an engineered shoreline of peninsulas, harbors, and beaches (figure 4.1). A reservoir of sand and clay suitable for use as fill was readily available on the lake bottom and on adjacent land. Furthermore, the natural lake bottom sloped gently from the shoreline to several thousand feet offshore, which made it feasible to fill the shallow lakeshore and build new land. In early 2008, the overall volume of lakefill totaled an estimated 57 cubic yards (52 million cubic meters) of material. The restoration of Montrose Point—a research and demonstration project that was created through a partnership between the Chicago Park District, the Lincoln Park Advisory Council, and the U.S. Forest Service—is an important attempt to restore pieces of land with surprising events via community involvement. Montrose Point is a peninsula on the North Side of Chicago that, like all the land along the shore of Chicago, did not exist before the 1860s.

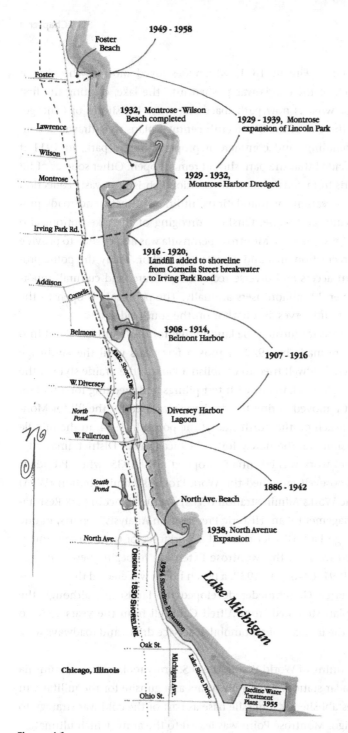

1949 - 1958

Foster
Beach

Foster

**1932, Montrose - Wilson
Beach completed**

**1929 - 1939, Montrose
expansion of Lincoln Park**

Lawrence

Wilson

Montrose

**1929 - 1932,
Montrose Harbor Dredged**

Irving Park Rd.

**1916 - 1920,
Landfill added to shoreline
from Cornelia Street breakwater
to Irving Park Road**

Addison

Cornelia

Belmont

**1908 - 1914,
Belmont Harbor**

1907 - 1916

Lake Shore Dr.

W. Diversey

North
Pond

Diversey Harbor
Lagoon

W. Fullerton

South
Pond

1886 - 1942

North Ave. Beach

North Ave.

**1938, North Avenue
Expansion**

ORIGINAL 1830 SHORELINE

1934 Shoreline Expansion

Michigan Ave.

Lake Shore Drive

Lake Michigan

Oak St.

Chicago, Illinois

Ohio St.

Jardine Water
Treatment
Plant 1955

Figure 4.1

The changing North Coast of Chicago. The dotted line in the middle indicates the
original coastline of the 1830s.

After the Great Fire in 1871, which destroyed most of the city of Chicago, much of the debris was pushed into the lake, creating the first load of fill for what is now park space along the lakeshores of Chicago. Although by the end of the nineteenth century, plans were drawn to construct large buildings and commercial property in the park, in 1911 it finally was decided that the park should remain open. Other sources of fill included debris from construction, demolition debris, and waste collection in the city. Excavation for foundations, utilities, tunnels, and roads produced clay, sand, and stone. Offshore dredging of the lake bottom also furnished fill for some land. Montrose peninsula was also created to provide more public recreation area and a protected harbor. Today the point provides lakefront access and diverse recreation, leisure, and cultural opportunities for over 20 million users annually. The peninsula is open to the north and east and serves as a harbor on the south side.

Although plans for turning the landfill that formed Montrose Point into park space were made by 1929, it took a few years until the landscape architect Alfred Caldwell tried to establish a park in the prairie style of the Midwest. In 1929, work began with test pilings and dredging for a harbor. The sediment removed during the dredging was used for the fill for Montrose-Wilson beach on the north side of the point. In 1934, in the middle of the depression era, the newly formed Chicago Park District amassed a huge debt, and work was essentially stopped until 1935, when President Franklin D. Roosevelt launched the Work Progress Administration (WPA) and the Public Works Administration (PWA). As the "Lincoln Park Restoration and Management Plan: Historic Preservation Analysis" reports, in one day the Chicago Park District received a workforce of ten thousand men, which allowed work on the Montrose Extension to begin anew (Chicago Park District 1991, 64–65). In 1937, a beach house was erected that remains there today. Ernst G. Schroeder developed the landscape, although the prairie-style plan stemmed from Alfred Caldwell from the years 1936 to 1938. Due to the increase of automobiles, service drives and roadways were created.

At the beginning of World War II, the U.S. Army took over the peninsula for use as a radar station. After a few years as a gunsite for the military, in 1956 it was established as a missile base as part of the cold war strategy to protect Chicago. Montrose Point was leased to the army, which ultimately used it as a Nike missile site (named for the Greek goddess of victory), housed three hundred soldiers there, and erected barracks, mess hall, and other facilities. By the 1950s, Montrose Peninsula extended over one kilometer into the lake compared to the prelakefill shoreline. Since at the time Nike missiles had a relatively short range, launch sites such as the one at Montrose Point were generally placed close to urban and industrial centers for protection from a possible Soviet strike. In 1960, Montrose's military area was surrounded by a security fence, which soon was covered by a hedge of nonnative honeysuckle bushes. Although the Soviet invasion never came, there was an unexpected invasion of another kind. Birds discovered the bushes and used them as a break area during their migratory journeys.

When the military left the peninsula in 1969, the surrounding fence was also removed, but remnants of the hedge, including the honeysuckle bushes, remained. By the early 1970s, when the area was reclaimed as a park space, the landscape consisted mainly of scattered trees and the honeysuckle hedgerow. In the early 1990s, plans were developed to restore the point to a savanna as it might have looked in the early nineteenth century. The old Caldwell plans were also referred to as an option for designing the park. The ecological reference point for restoring Montrose Point in the literal sense of the word would be to turn it back into part of Lake Michigan. Instead, ecological restorationists are attempting to design a landscape as it might have looked before European settlement—if the land that is now Montrose Point had actually existed and the savannas of Illinois had reached into Lake Michigan.

Public Participation and Controversies over "Real" Nature

An important development in the north coast of Chicago started in the spring of 1996. In preceding restoration projects in the metropolitan Chicago forest preserves, a controversy erupted over the ecological restoration of prairies and oak savannas (cf. Gobster 1997; Ross 1997; Siewers 1998). The local and even national news took great interest in what came to be known as the "Chicago controversy." This controversy grew from arguments about the objectives of restoration in suburban communities in Chicago. Opposing groups were successful in interrupting most of the restoration activity in the Chicago area for about a year. This controversy was used as a touchstone to examine questions about peoples' values associated with the unpredictability of nature. As Maria Alario (2000, 496) points out: "The controversy surrounding the Cook County Forest Preserve project provides particular insight, as restorationists are coming to realize that knowledge, financial resources and even extensive volunteer work are no substitute for democratic participation in decision-making process."

These conflicts have been instructive for the city of Chicago and the future involvement of nonscientific experts. They were a surprise to the stakeholders, but they "have been beneficial inasmuch as they have publicized how passionately people feel about which strategies are adopted in efforts to preserve the natural environment" (Alario 2000, 490). Most people did not object to the idea of restoration but centered their concerns on specific practices, such as tree removal, the killing of deer, and especially the timing and methods of these events. The split was between restoration efforts and traditional conservation and preservationist ideas. For citizens who took a preservationist stance, the main goal was not to restore but to preserve what was left and what was there, whether this was rare native or beautiful nonnative flora and fauna. The restorationists saw their most important challenge to be to bring back native prairies and savannas, whose native plants and animals where about to drop out of the gene pool.

In both cases, the wider public simply wanted to have a say and even to get involved in the process. The United States Department of Agriculture (USDA) and the City of Chicago thus learned from this experience that the involvement of the wider public can be the key to successful restoration strategies. In the subsequent development of Montrose Point, a bottom-up approach was confirmed by the City of Chicago as being the most successful strategy. The Nature along the Lake program was initiated to involve the wider public to provide lakefront park experiences. It was customized to school curricula so that students learn about the lake as a habitat system, birds and their migratory patterns, aquatic life, native plants and trees, as well as other outdoor science topics. Furthermore, a Friends of the Park Program (FOTP), which in an earlier form was founded in 1975, is providing nature experiences and environmental education to Chicago public school children at Montrose Point along the beach. It is a park advocacy organization that is dedicated to designing, preserving, and improving Chicago's parks and forest preserves for all citizens.[6] Restoration workdays are advertised and announced in local newspapers, flyers in the mail, and other media. They are held throughout the year and include activities such as seed collection and planting, brush clearing, and invasive species removal.

This is a style of public knowledge production where a key factor is the engagement of participants as knowledgeable agents who are capable of dealing with experimental practices and surprising turns. Today community involvement in general is regarded as the key to a long-term survival of Chicago's nature preserves. Since 1996, as Kathy Dickhut from the Lincoln Park Advisory Council (LPAC) stated (as quoted in Furnweger 1997, 3), the situation is very different: "Whoever wants to be involved in the process can be. . . . The results will then be given to the landscape architect we hire to inform the design process. And the plan that person comes up with will be subject to a lot of review by the planning committee, which will also have community people involved."

Although community remains an important symbol and aspiration in political decision making, the word does not imply a romantic idea of harmonious entities that foster learning processes to improve a common situation and take action collectively (Brint 2001a). Instead, it often is something you have to do, as Kathy Dickhut states in a 2008 interview: "Whether you want it or not, you have the community process. Even if you don't want to do it [participation], it's necessary." The long and laborious negotiation processes that took place surrounding the work that was done at Montrose Point expose complex organizations that are made up of people who sometimes act unreasonably, inject their own values, and use forms of cooperation and power to pursue their own goals.

Given that restoration at times can be seen as a transdisciplinary type of science, its societal context needs to implement the unexpected consequences of interactions. Thus, questions arise about how social and "scientific" factors can be knitted together so that the wider public's knowledge can become a meaningful element in restoration ecology. Members of the public do not have expertise in every specialist domain, although they might have a vast store of what Collins and Evans (2007) called "primary source knowledge" and, more important for restoration practice, "interactional expertise." Closely interacting with the experts accomplishes the transition from primary source knowledge to interactional expertise—that is, kinds of knowledge that come from studying primary literature or from pure observation. Interactional expertise is slowly gained with more and more discussion of the science. This type of knowledge is thus produced out of the direct relationships between people and their "natural" environments. With the right dose of knowledge and interactional expertise, the so-called nonexpert can become part of restoration work and science and can even contribute to academic science.

This means that an initial observation by an expert, which can be seen as a starting point of a project, can go hand in hand with the problems that are perceived by people who are not involved and not detected as experts. Experts here can as well mean lay experts, stakeholder groups, or academic ecologists. Indeed, the first ideas about nature came from the "uncertified lay experts" (Collins and Evans 2007) of the surrounding communities, and the academic ecologists were later invited to devise (together with the stakeholders involved) solutions that would be ecologically feasible and desirable. Included in the stakeholders were representatives of the adjacent communities, organized interest groups (such as the anglers, the boating club, and the volleyball group), and users of the park who were randomly selected on several days as they visited the park. This increased the public's acceptance of the negotiations since the procedure suggested that almost any park user might have been invited to participate. Scientific experts and the broader public met by evaluating options, negotiating strategies and adjusting knowledge before an intervention, since "restoration ecology is both an experimental science and an urban project, where each step must be subject to review and debate" (Alario 2000, 497).

Several analysts have noted that participation should follow an appropriately "democratic" process in terms of participants and agenda. In addition, greater openness and transparency have been called for on many occasions in research on decision making in administration. Most generally, openness in environmental decision making is regarded as part of "good governance" and often involves greater transparency in organizational structure and wider participation in some or all elements of a decision and implementation process (cf. Cooke and Kothari 2001; Hood and Rothstein 2001; Mol 2008). Many different models of public participation and the resolution of environmental conflicts exist and have been criticized (e.g., Abels and Bora 2004; Agrawal and Gibson 2001; Rauschmayer and Wittmer 2006; Smith 2008). The experimental approach developed here aims to open the door to a new way of understanding ecological design and participation as part of one "experimental cycle." Other forms

Table 4.1

A typology of participation

Type of Participation	Description
Coercing	The will of one group is imposed on another group. People passively adapt to a decision with little ability to change it.
Informing	Information is transferred in a one-way direction. Participation as seen as a strategy to inform people. Is usually done after major decisions and implementations have already been made
Consulting	External agents are hired to ask questions related to the planning of a project.
Colearning	Different expert groups work together to share different sets of knowledge to make decisions. Also called an interactive approach.
Coaching	Knowledge is shared and sometimes produced in situ between the different groups engaged. Also called the self-mobilization approach, where people set their own agenda.

Sources: Extracted from Arnstein (1969), Keen, Brown, and Dyball (2005), and Pretty (1995).

of participation in relation to the ones involved at Montrose Point can be seen in table 4.1. The first four types of participation become more important in the following chapter on the design of postmining landscapes in the southern parts of Leipzig. Colearning and coacting describe the participation that is going on at Montrose Point, whereas the other forms of participation information transferred in a one-way direction have more relevance in the case discussed in chapter 5. This should mean that coacting and colearning are at the core of a participatory process that helps people deal with surprising events. Because people get involved in analysis and conditions of action plans, they need to be ready to engage and mobilize themselves, initiate actions, and take responsibility for the surprising side effects of their actions.

Surprises Appropriated: Native Birds, Nonnative Bushes, and the Arrival of Baby Dunes

The general starting point in restoring Montrose Point was that residents wanted their nearby park areas to be more natural. Through the participatory process, different stakeholder groups worked to negotiate a design for

Montrose Point. Paul Gobster (2001) categorized four main visions of what Montrose Point stakeholders regarded as "natural" by analyzing focus-group interviews and debates in workshops beginning in 1997. The focus-group sessions with diverse stakeholder groups were conducted to arrive at a vision of what nature means for the Point; how such a natural area should be used for recreation, education; and how this vision can be realized through physical design, management, and programming. From 1997 to 1999, a strong consensus emerged among groups that wanted to enhance the natural qualities of the site, to respect the intent and integrity of the Caldwell plan from the late 1930s, to provide for use of the site, to provide appropriate access through the site to popular fishing and volleyball areas along the lake, and to develop appropriate maintenance and management programs.

Although the goals that were agreed on (such as nature enhancement, historic restoration, and appropriate recreation use) are compatible with one another on a broad level, translating them into specifics has raised a number of significant questions. One crucial question, for instance, was whether a "real" prairie with a diverse mix of species, heights, colors, and textures could actually substitute for the mowed lawn that was characteristic of the Prairie Style meadow advocated by Caldwell and others. A further crucial challenge was that "wild" landscaping could also create and even foster unintended uses and perceptions of the area, although they can be communicated only as knowledge about the unknown. Thus, the purpose of the workshops was to examine the integration of historic-preservation and nature-enhancement goals in the restoration of the park. Findings from the workshops were later used to inform planning, design, and management alternatives for the Point. The workshops brought Chicago Park District staff and public-interest group representatives currently working on the Montrose Point Restoration Project together with experts in historic preservation, ecological restoration, and other areas to help resolve specific issues that were identified in the concept-planning phase of the project. The first plan that was agreed on blended different ecological and recreation concerns into a prairie-style site design since many competing recreation interests existed among the interest groups.

However, the hedge in the middle of the peninsula, which developed from the remnants of the military hedge, caused some stir. The hedge consisted mainly of honeysuckle hedgerow, an invasive and nonnative plant from Europe, and the plan to restore Montrose Point to a native savanna was questioned since the hedge also attracted many native birds, some of them unseen in Illinois since at least the 1940s. By 1977, the birdwatcher community had baptized the row of the honeysuckle bushes the "Magic Hedge" because the hedge was a magnet for birds even in an area with many visitors, hikers, picnickers, and joggers and a busy road close by. The name and the birds became better known in the wider public as well as among academic ornithologists beginning in the early 1990s.

Today the Magic Hedge is made up of woody plant species that range from red maple to the traditional honeysuckle.[7] The area between the Magic Hedge and Lake Michigan comprises the majority of Montrose Point and is dominated by weedy forbs and grasses mostly of Eurasian origin (cf. figure 4.2). After cold weather in fall or bad weather in spring, the hedge often harbors an immense number of birds. Birds migrating north and south through Central and North America are funneled through the Chicago area by Lake Michigan. Montrose Point, which projects into the lake can thus be called a natural, albeit human-made resting place for

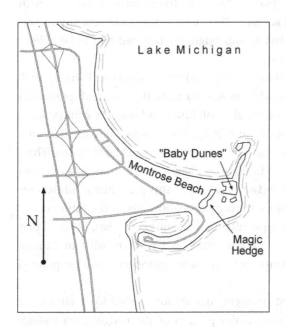

Figure 4.2
The Montrose Peninsula with the Magic Hedge in the middle and the location of the baby dunes on the north beach. (Courtesy of C. Kuhlicke)

migrating birds. When randomly picked interview respondents at Montrose were asked what they liked most about the point, they most frequently mentioned the Magic Hedge (n = 15, 40 percent), especially its relevance as a bird habitat. As one visitor explained: "I like the ruggedness of nature, and the wildlife and birds you see" (Feldman 2007, 88). The fact that these birds were attracted to the hedge and voluntarily rested there was repeatedly communicated (even by local media and academic ecologists) as a surprise, especially after the controversy in 1996. At a focus-group meeting, one participant aptly remarked that "we can do the monitoring, but birds, I don't know. If you read the paper, you'll see there's no real rhyme or reason sometimes about why they use what they use." Furthermore, it was relatively unknown territory for academic ornithology. Greenberg observed that in general "very little research has been conducted on migrant birds—their requirements during this stressful stage of their lives are poorly known" (2002, 400), "perhaps because most professional ornithologists are fearful that such an inquiry would smack too much of birding than of science" (ibid., 407). The observation of migrating birds diluted the original plan to "restore" the point to a savanna with mainly native plants and animals. The exotic honeysuckle eventually was determined to be not as bad as was originally believed and to be able to enhance local environments.

The conflicting interests in the "restoration" of Montrose Point in 1999 led the human actors involved to revise their goals, theories, and approaches according to the location's natural possibilities and people's cultural ideas about nature. New ecological knowledge about native birds and resting habits led to a strategy change that now protects the Magic Hedge. Thus, an intervention, as depicted in figure 4.3 below, can also mean a decision for nonintervention. The decision that led to this particular nonintervention was easily agreed to since during negotiations about the future of the hedge, some stakeholders, especially the academic ecologists, believed that the Magic Hedge would disappear within a few years without human intervention as native plants and trees were introduced and supplanted the hedge.

In 1998, some ecologists thought that aphids seemed to be killing the honeysuckle that makes up a major portion of the hedge. Even though users (especially the birdwatcher community) wanted the hedge to remain undisturbed, some ecologists suggested replanting it with mainly native

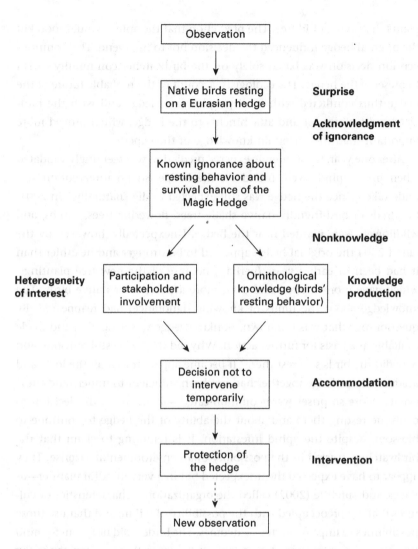

Figure 4.3
From a surprise to new knowledge for decision making

plants that were aphid free. The prediction that the aphids would soon kill the original hedge influenced the decision not to intervene. The nonintervention decision was based solely on the birdwatcher community's wish to preserve the hedge. The ecologists' account of the probable future of the hedge thus conflicted with local people's knowledge and with the birdwatchers' feelings for and attachments to the hedge, which proved more important than the uncertain knowledge of the experts.

After one year, the academic experts' prediction was seemingly validated when more aphids were found on the hedge, so no intervention was undertaken since the hedge was still expected to die "naturally." In 2001, hundreds of bird-friendly native shade trees, flowering trees, shrubs, and wildflowers were planted near the hedge. Unexpectedly, however, by the fall of 2004 the original hedge appeared to be stronger and healthier than it had been in earlier years (perhaps because of the new tree plantings, which gave more shelter) and even more attractive for native birds. The knowledge about the limits of knowing (ignorance) had reappeared. By questioning what was unknown, nonknowledge was generated and made available as a basis for further action. Why did the hedge still blossom, and why did the birds still rest there? It became apparent that all the local and academic knowledge together had not been sufficient to understand these events. More surprises were communicated—about the birds' decision to continue resting there and about the ability of the hedge to continue to blossom, despite the aphid infestation. It is tempting to claim that the birdwatchers trusted in the possibility of an environmental surprise. They appear to have expected the unexpected but in a very mindful manner—as Weick and Sutcliffe (2007) called the organizational characteristics of cultures that are preoccupied with the possibility of failure and that use those possibilities to improve system outcomes. The birders did not seem to mind that the new restoration design included a space that was ten times the size of the hedge. They wanted to keep their original hedge, which had almost "attained an iconic, even sacred, status" (Gobster 2004, 8).

Another surprising observation temporarily overlapped with the negotiations over the Magic Hedge in the fall of 1999. A volunteer noticed some striking green stems that were emerging from the sand near the shoreline on the north side of the point. In the following May, the plant was identified as lakeshore rush (*Juncus balticus*), a plant that had not been found on Chicago's lakefront since 1946. The lakeshore rush was growing on one of two dune formations that had begun to develop a few years earlier. As usual in nature, the two "baby dunes"[8] began with a coincidence, in this case probably via an oversight by human beings. Members of the Chicago Park District might have neglected to properly comb a section of the sand, and then, as Noreen Ahmed-Ullah of the *Chicago Tribune* reports, "nature took its course—no matter that it was a man-made beach in the first place" (Ahmed-Ullah 2007, 5). Between 2006 and 2008, the dunes have doubled in size, and they continue to grow.

One could say that nature had expanded its reach onto a formerly public beach. Together with other plants, the rushes were helping to bind sand, which would lead to further dune development. The dune expanded when the Chicago Park District stopped grooming that section of the beach, and after a few seasons vegetation spontaneously took hold, including one plant that was an Illinois state endangered species. The beach dunes are a 9.25-acre (3.75-hectare) site that is located at the east end of Montrose Beach (Chicago's largest beach). In October 2005, the Illinois Nature Preserves Commission added them to the Illinois Natural Areas Inventory (INAI), a statewide list of high-quality natural areas (the first time a Chicago Park District parcel was so recognized). The dunes are being restored to their "natural state" by the Park District and several volunteer groups that were instrumental in setting aside the site as a natural area in 2000 (see figure 4.2). The goal of the restoration is to encourage the native grasses, sedges, rushes, and forbs adapted to beach, dunes, and swales that are increasingly colonizing the area. The volunteers also monitor several state-listed plant species for the Chicago Botanic Garden's Plants of Concern Program.[9]

In general, the local newspaper articles that reported on the development of the dunes between 2002 and 2008 used terms (such as *surprise*, *mysterious*, and *strange*) that indicated that the dunes were perceived as a surprise. As the Park District's natural areas manager, Zhanna Yermakov, noted: "The fact that these dunes form naturally in the middle of the city is remarkable. This is not off the beaten path. This is Montrose Beach" (Ahmed-Ullah 2007, 5). Dunes are dynamic and constantly moving, and their continuous changes can easily be observed by hikers, volleyballers, and everybody passing by regularly. In referring to the arrival of the first baby dune on Montrose Beach, Ahmed-Ullah (2007, 5) sums up: "Just as it came mysteriously, it could easily be eroded by the weather or submerged under rising lake levels. Or it could get bigger, though officials are trying to keep it from growing over a path to the south." For Debra Nelson, the Chicago Park District's natural biologist, "How long it lasts will be up to mother nature."

This surprise at the development of the dunes did not lead to dismay, although the original plans had to change, which meant a major shift in how some actors (such as the volleyball interest group) perceived nature. The Montrose Beach volleyballers initially feared that one day the baby dunes would grow up and "roll over" their volleyball area. The volleyball group initially objected to the growth of the dunes and pled for an intervention that would stop the growth. However, they soon discovered that dunes with bushes provided protection from the stares of passers-by. Although no one told me officially, I also noticed that the volleyballers used the bushes on the dunes as a "natural" comfort station. Here is another example of unexpected positive outcomes from an event that was originally perceived as negative. If the dunes continue to grow, however, the current volleyball field may have to be moved northward. Overall, the visions of various interest groups are being integrated with the natural development of the dunes and the revised plans for Montrose Point.

Maintaining Integrity in the Face of Surprises

After more than ten years of restoration, Montrose Point is at the point where the different visions of nature of all the interest groups (including natural scientists and landscape architects) have been shaping the design of the area. Via their work at the Point, the natural scientists and volunteers have developed new forms of unique local knowledge that are important for several areas of science. Spurred by the observations of lay birders, ornithologists are studying migration routes. Community-led restoration activities have also stimulated some academic ecologists to study migration patterns as well as other ornithological issues at Montrose Point.

To learn more about these sources of knowledge about native birds, focus groups have been held to help inform the design and management of each part of the site (Gobster 2001). These groups have included birders, restorationists, historic preservationists, recreationists, anglers, people with disabilities, and members of neighborhood communities. In this bottom-up research collaboration, new research questions and new scientific challenges have evolved over time via continuous interactions between different groups (cf. Eden and Tunstall 2006; Küffer 2006; Pohl and Hirsch Hadorn 2007). For instance, without the focus-group meetings and negotiations between different stakeholder groups, birdwatchers would not have shared their knowledge with other groups, and academic ecologists and ornithologists could not have appreciated the birders' knowledge and contributions to their respective fields. The birdwatchers' regular field observations cannot be easily expressed in propositional terms, which is a typical problem with many kinds of lay knowledge: the knowledge of the birdwatchers cannot be easily represented by other individuals but must be articulated by the lay experts themselves.

Spending time in an ecological setting helps to develop the so-called tacit knowledge (cf. Polanyi 1958) that can be important for basic research. Not only have the birdwatchers developed a special knowledge through firsthand experiences, but they also are commited to supporting the long-term development of the area and are highly motivated to learn about its ecosystem. The stakeholders are encouraged to learn about various interest groups, since learning is perceived to be personally important, and the professionals asked pertinent questions about topics that were outside their immediate area of expertise, which enabled them to uncover and translate the implicit knowledge of the birdwatchers' long-term observations and experience.

Without the debates surrounding the Magic Hedge, perhaps the Chicago Ornithological Society and the Audubon Society would not have supported the preservation and enhancement of the hedge. In the years since the Chicago Park District's 1996 designation of Montrose Point as a nature sanctuary, the District's Nature Areas Program has focused on providing habitats for migratory birds in the parks that line the city's shoreline. Based on recent research, the city has installed a system of bird sanctuaries at strategic intervals along the lakefront and beneath the hedges at Montrose Point. These sanctuaries, which Paul Gobster has characterized as "fast-food stopovers," provide places where birds can rest and feed on berries or insects before heading back out on the flyway.

In short, without an experimental engagement that unfolded natural dynamics *and* social changes, there would have been no successful restoration of Montrose Point. By sharing their experiences, expertise, and different sources of knowledge, the stakeholders developed a unique type of knowledge. The design of the landscape at Montrose Point can be understood as an example of achieving "communion with nature" (cf. Jordan 2003) through an experimental interaction between nature and the human community.

The dynamic nature of Montrose Point means that the responsible actors involved are constantly negotiating with developments within the ecosystem, which must be part of any successful ecological design project. The focus-group meetings clarified the unknowns and uncertainties that the public faces when deciding whether to adopt a particular strategy. Although the various interest groups had their particular interests and emphases for each site, most individuals agreed on the uniqueness of the places and the complementary values that could be achieved through management. The dialog has helped to construct robust ecological designs that are suited to each area and are based on negotiated considerations about desired naturalness, degree of human access, and amount of public participation in planning and management (Gobster and Barro 2000; Martin 2005).

The project at Montrose Point continues to evolve. It is moving out of one phase in response to changes in actors, policies, and natural factors (observation, surprise, knowledge production, articulation of interests, potential accommodation) into a new implementation. The overall integrity of the restoration process at Montrose Point has been maintained because its design has accommodated revisions and modifications as issues that were previously agreed on ended up changing. In this recursive process, the actors experimented until they found a solution that appeared to be context sensitive. The process cycle is closed since when the outcome of the activities is evaluation, the evaluation is applied to previous assumptions (hypotheses), which in turn are accommodated where necessary. The actors involved therefore were constantly negotiating with the material world and the social world to meet internally set goals. It can be a rather playful process in which steps into the unknown are taken and things are tried out. Furthermore, the strategy has allowed steps to be taken in a flexible process in which stakeholder participation leads to accommodations in the implementation of the restoration plan in the next phase without undermining the principal goal of the plan—to design a more "natural" park. In this process, the linkages among research, implementation, and social acceptance have become more robust.

Beginning in about 2006, a few changes became apparent. Some of the stewardship component of the restoration at Montrose Point has changed since the 1990s. There have been two new natural area directors at the Chicago Park District since 2001, and each changed some institutional memory. Some verbal agreements that were made among the stakeholders have been lost. So far, the birders and other groups still seem to be satisfied, and there has not been much conflict (cf. Feldman 2007). Paul Gobster and other actors seem to be worried, however, that the city is planning new projects in a vacuum. The birdwatching community is already indicating that it will soon protest the projects. Quite a few volunteers protested against a fence that was built without any public participation process. In reaction, the park's spokeswoman is cited to have said: "We certainly welcome the feedback of our volunteers, but ultimately we must implement what we feel is the best management plan for our nature area" (cited in Ahmed-Ullah 2008, 4). Unfortunately, this approach runs contrary to many of the things that have been crucial prerequisites for successful public ecology projects in Chicago in the past decade.

Aligning Research and Heterogeneous Social Goals

In ecological restoration, a volunteer approach can be important for academic research for several reasons. Whereas conservation and preservation generally focus on the threat of species loss, restoration focuses on a long-term re-creation or even invention of landscapes. Practitioners believe that one reason that public restoration experiments run mainly by concerned citizens are successful is that practitioners see them as long-term enterprises. Ecological restoration is unlike other traditional sciences since it is based on trying out things even if the knowledge available is miniscule. It thus is a means for launching an ecological design project in spite of ignorance and for upholding it without disrupting the overall process. This type of technique was rejected by many traditionally oriented scientists. Recalling some of his critics in the early days of ecological restoration in the late 1970s and the 1980s, Stephen Packard says: "The idea that someone thought they might be able to learn something new about a revered natural community through lowly restoration experiments seemed especially to offend these critics" (Packard 1988, 18). Many of the academic ecological scientists did not regard this as a reliable source of scientific knowledge, as has been discussed above.

Restoration does not follow a fixed master plan of action. As the restoration of Montrose Point shows, the plan needs to be pieced together and built, thought about and tried out, formulated and reformulated—always in negotiation with other people *and* the natural world. Distinctive features of ecological restoration's performance are that it is experimental, inherently uncertain, and based on learning by doing (cf. Covington 2003; Walters and Holling 1990). It is a site-specific approach that fully considers the surprising reactions of the nonhuman nature that is found in these sites.

The restoration practice at Montrose Point depends on what we have called *nonknowledge* as well as *negative knowledge*. Negative knowledge points to the limits of knowing, but it also refers to the things that we want to know (although we do not necessary like them). Stephen Packard said, "Let the fire decide" about the prairie burns that have been conducted in the Chicago area since the late 1970s (Packard 1988, 14). This is not unique to Chicago. Practitioners in a program to protect endangered fish in western Colorado have agreed to monitor the response of fish populations to certain recovery actions to "let the fish tell us" whether those actions are sufficient to meet recovery actions (Coe-Juell 2005). In other words, ultimate knowledge of how nature acts remains outside the grasp of the scientific ecologists, volunteers, and citizens who are involved in restoration. As new problems keep appearing, restorationists continually have to explore nature in their work in a process of "reciprocal tuning" between people and the natural world. Significantly, this tuning occurs within a trajectory toward a new alignment in which each "partner" interactively sustains each other. The importance of this approach lies in the fact that the basis for a traditional scientific (that is, foreseeable) decision cannot be known. In many complex situations, ecologists cannot say exactly what types of measures are needed to react to certain natural events. In other words, vigilant and clever restoration practitioners knowingly admit the limits of knowing. This can be connected to Paul Gobster's (2004) usage of Andrew Light's (2003) notion of "green time" to point to the difference between human and (often unknown) ecological time and to allow nature to unfold without stopping to set goals for desired future conditions for the human community.[10]

The case of Montrose Point has shown that the expectation of "surprises as opportunity" merged into a time scale that was "greener" than the ecosystem that some actors planned to design, but it also rendered obsolete the idea of a detached and austere ecological science. Surprises can be perceived by different actors, who then initiate new processes of scientific practice that veer toward previously unimagined goals. This calls for a science that is robust enough to listen to both different interest constellations and unexpected natural changes. In this sense, knowledge production in the real world beyond the laboratory must be able to embed the learning process in a way that allows new surprises to be absorbed with fewer problems than traditional management strategies of scientific implementation introduce.

Such surprise-welcoming experiments in the real world often become part of the public's everyday life, so negotiation of the design among the heterogeneous actors can be accomplished by turning concerned citizens into hands-on practitioners to handle surprises and to renegotiate the design. The restoration of Montrose Point has shown that conducting time-consuming and laborious hearings and involving volunteer group organizations and stakeholders via focus groups have lead to more robust strategies in the long run. The original project design at Montrose has been repeatedly refined to account for the frequently changing ecological *and* social realities. This can be called a matter of "uncertainty absorption," which differs from the concept put forth by March and Simon (1958, 165), since it is a question not of avoiding or absorbing uncertainty and risks but of coping with surprises that are anticipated (and possibly welcomed) and surprises that are not anticipated but nevertheless dealt with appropriately. This connects to what Stephen Collier and Andrew Lakoff (2008) call the rationality of preparedness. They propose that the objects to be known point not to issues such as economic incentives or a population's health issues but to issues such as a robust infrastructure that provides the continuity of social order. Besides the access to certain financial resources, a regional administration needs to support an experimental approach with an efficient administration (for example, for making quick adjustments to unexpected events). This, however, shows that an experimental mode of knowledge production in the real world has nothing to do with romantic or even essentialist notions of bottom-up grassroots activities. Successful bottom-up approaches, quite paradoxically, appear to need even more tightly and carefully planned legal, financial, and organizational frameworks by decision makers and policymakers.

In addition to the factors mentioned above, a "critical infrastructure" is needed. This means that negotiations between stakeholders need to be characterized by flexible decision-making procedures in case of surprising observations. However, a community that wants to be involved in a restoration project also needs to be flexibly engaged in knowledge production that is relevant to shaping their lives and environments. This is not self-evident. In a political climate that fosters the belief in certainty through science, public experiments as outlined here will not work. Ecological design projects are often undertaken on tight budgets, and they cannot be postponed until every detail of future planning is precisely known.

Before an experiment can be successfully launched, the community needs to be aware that no action is action as well. If the members of the community wait for an ultimate scientific truth, they are misusing science as a source for political nonaction. However, choosing not to act also leaves surprising effects in its wake. If science proposes to deliver reliable knowledge, then it overestimates its methodological competence. As Naomi Oreskes (2004) argues, science cannot and does not generate certainty; at best, it is able to produce useful processes of inquiry that allow for a continuation of further examination and a revision of earlier results (cf. Sarewitz, Pielke, and Byerly 2000). The way out of this dilemma appears to be a mutually agreed on strategy of experimental practice so that discovering and observing unexpected effects do not suggest failure. The experimental approach should allow participating community members to be aware that ecological strategies are always based on a serious amount of ignorance. However, ignorance as an experimental strategy can comprise alternate phases of corroborating the robustness of a strategy by practice. Ideally (and the example of Montrose Point supports this ideal type), the result can lead to more reliable knowledge (such as expertise about the resting places of migrating birds), which can in turn be used to produce new or extended knowledge (cf. chapter 3). Ecological practice thus can embed the learning process in a way that allows new surprises to be absorbed without bringing the project to a halt. Duncan (1999) has suggested that in a climate where the public seeks certainty, a way forward could be that a community that supports the decision to act in face of ignorance should be rewarded in a form of compensation.

Since heterogeneous actors (such as birdwatchers, hikers, and ecologists) can participate with many different interests and knowledge bases, the certified experts (that is, experts with PhDs in botany or biology) are much more accountable for the results than they would be in their traditional ivory tower research. In traditional research, virtually the only control mechanism is internal peer review. When actors with lay knowledge, experience-based expertise, and other "alternative" modes of knowledge production are embedded into restoration activities, the peer-review process is extended to the wider public (cf. Funtowicz and Ravetz 1990). Nonacademic research thus becomes a pivotal part of a more encompassing activity of ecological practice, but it occurs in close cooperation with traditional knowledge production (mode 1) and without undermining the decisive character of science, which also includes the fostering of surprises and its attempts to subsequently control them.

Bruno Latour (1999, 91–92) argues that the traditional model of science resembles a core surrounded by a corona of social contexts that are irrelevant to the definition of science. In his view, a temporal shift has always been taking place in science, and the whole internal/external question needs to be reconfigured. In this view, successive chains of translations occur, and through this work actors modify, displace, or shift various and sometimes contradictory goals and interests that involve "exoteric resources" (including information from the newspapers, television, and nonscientific goals such as aesthetics and community well-being). There are also "esoteric resources"—that is, internal scientific information from scientific journals and the like. As Latour puts it: "Everything important happens *between the two*, and the same explanations serve to carry the translation in both directions" (Latour 1999, 91). Important debates have circulated around these issues, but it is not always clear whether they allude to changing types of resources or to the epistemological foundations of science. In ecological restoration, it seems that esoteric mode 1 science cannot make progress without exoteric mode 2 science, and vice versa. The restoration of Montrose Point has integrated important components of both sides. In other words, social robustness appears to go hand in hand with a high epistemic robustness. Knitting together scientific reliability and

its social counterpart provides a way of building a robust working relationship between ecologists and the larger community and to foster ecological
surprises in restoration as a powerful way to test basic ideas about the
ecosystem being restored.

Robust Restoration Strategies through Recursive Practice

Notions of robustness exist in many disciplines, including engineering,
computer science, and operations research, and they often are contradictory. The widely discussed book *Rethinking Science* (Nowotny, Scott, and
Gibbons 2001) introduced the term *robust* to the field of science studies by
asking how science can be socially embedded to allow its implementation
to be "socially robust" (cf. chapter 2). Regarding socially robust science,
Helga Nowotny and her colleagues (2001, 54) state that "contextualization
is invading the private world of science, penetrating to its epistemological
roots as well as its everyday practices" and claim that social robustness
leads to a condition where science's "epistemological core is empty" (179).
The consequence of this change would be that science is now moving
beyond its classic ideal of producing reliable and thus true knowledge to
a new mode of generating socially robust knowledge.

Commentators, most forcefully probably Peter Weingart (e.g., 1997,
2008), have argued that claims like these lack empirical foundation and
do not have any terminological clarity. In a way, the term *socially robust
knowledge* as introduced by Nowotny and colleagues appears to be an oxymoron. What could it possibly mean? Judging from the writings of Nowotny
and others, it should mean a type of knowledge that is valid across scale
and time independent of social influences. But Nowotny et al. sometimes
understand the term *social* as an indicator for a more sensitive and site-
specific form of scientific knowledge that is able to incorporate "the 'soft'
individual, social and cultural visions of science" (Nowotny et al. 2001,
198). How can this second type of site-specific socially robust knowledge
be robust in other contexts? If socially robust knowledge is local and thus
place bound, how can it be used to deliver reliable guidance for science
and knowledge production in other regions with site-specific social frames
and contextualizations? Does socially robust knowledge imply an epistemic change in the production of ecological knowledge (what Nowotny
et al. sometimes seem to allude to), or is it merely a call for a more context

sensitive type of implementation (as they also indicate occasionally)? According to these two interpretations of the term, social robustness does not seem to be able to integrate both aspects in a meaningful way. In light of this, robustness thus does not seem to be a useful indicator for the quality of knowledge per se.

Robustness, however, does appear to be an apt characterization of a *process* that successfully deals with ignorance and surprise in ecological research and implementation strategies. This idea of robust processes comes closer to notions of robustness as used in research on technical and biological systems. Carlson and Doyle (2002, 2539), for instance, refer to robustness as "the maintenance of some desired system characteristics despite fluctuations in the behavior of its component parts or its environment." It can be related to Joy Zedler's (2007, 164) idea of successful ecosystem restoration, where a judgment of success is based on whether an ecosystem and its human inhabitants can sustain themselves without further maintenance and outside control.

Furthermore, the notion of robustness is both related to and different from the concept of resilience as used in discussions on adaptive management, where a social-ecological system is able to cope and adapt over time while essentially retaining its basic characteristics (e.g., Folke 2006; Gunderson and Holling 2002). Despite its many and often ambiguous meanings (cf. Brand and Jax 2007), Anderies, Janssen, and Ostrom (2004) have shown that resilience theory's emphasis on adaptive capacity can be useful in a descriptive sense. However, it is not applicable for the understanding of designed systems since it appears to be impossible to design for adaptive capacity. Robustness, in turn, emphasizes the tradeoffs that are associated with systems designed to cope with uncertainty, and thus it appears to be a more appropriate concept for trying to understand surprises, disruptions, and setbacks during the design process (cf. Anderies et al. 2004). Robust processes should be able to include directions of basic change by focusing on the strengthening of knowledge production for steering the development in such a way that the overall integrity of research *and* implementation can be upheld. Robust processes thus point to the possibility that a chosen course of implementation and development may be effective in face of changing external social and ecological conditions. In this sense, the term *robustness* points to two seemingly contradictory aspects— controllability and flexibility. Positioning surprise and ignorance as the cornerstones of public experimentation with an ecosystem, however, calls for the modularization of the control and a flexible reaction to surprising events. These can include surprising setbacks, radical revisions, and modifications to issues that were previously agreed on and then subject to change. Even Nowotny et al. (2001, 55) state in passing that the contextualization of science and knowledge production "can actually enhance scientific reliability"—although they do not explain how this can come about. In the rest of this section, I suggest how robust strategies can include a context sensitive approach and, at the same time, enhance scientific reliability.

We have seen that ecological restoration can be a new and challenging form of cooperation between different knowledge producers and ubiquitous types of experts. Robust strategies in ecological restoration can be faced as an emergent practice that is based on recursive processes that allow learning in and with the ecosystem (Berkes and Turner 2006; Keen, Brown, and Dyball 2005; O'Brien and McIvor 2007). Social expertise in ecological restoration should not be a substitute for ecological reliability since a methodology based on experimental practice can have the potential to strengthen scientific rationality *and* social acceptance. The restoration of Montrose Point can be understood as a form of research that tries not to juxtapose scientific importance and ecological science with the irrational and culturally tainted ideas of the public. Rather than undermining the science of ecology, the wider public has become an important part of ecological restoration and has moved beyond the traditional conflict between "ignorant" lay people and rational or objective natural scientists. People are not passive subjects but active agents in the scientific process. In ecological field practice, contestation of restoration knowledge (such as the resting habits of different birds) via the involvement of the human community has produced a transformed and enlarged definition of *scientific research* and has fostered a public involvement in science. The relative lack of control in boundary conditions can be absorbed and compensated by the recursive quality of the research process and the institutional steps in the design cycle. The recursive process allows both positive and negative experiences to be fed back into the next step of the design process. An informed consent about what is known and unknown appears to be a prerequisite to welcome surprises. This circumscription points to the paradox discussed before: whenever new knowledge arises, the perceivable amount of nonknowledge increases at least proportionally.

Figure 4.4 illustrates how the benefits and disadvantages of experimental strategies in the development of Lincoln Park and Montrose Point temporarily can vary with their scope of preparedness and openness for surprises.[11] The prior knowledge base and clear knowledge about what is not known (nonknowledge) constitutes a fourfold matrix that is related to the organizational level of openness to surprises. Successful ecological design projects require socially and scientifically robust research and application strategies. As is exemplified in the case of Montrose Point, these strategies have two major elements—openness to surprises and knowledge about what is not known (that is, ignorance turned into nonknowledge).

Acknowledging ignorance and being open to surprises means that the actors accept an experimental approach so that surprises are not judged and communicated as failures. It can be called a *safe-fail approach* or what

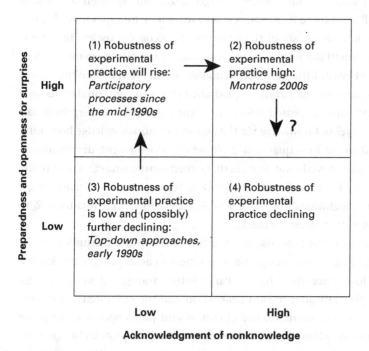

Figure 4.4
Ignorance and surprise in the dynamics of ecological design strategies

organizational theorists call "the strategy of small losses" (Sitkin 1995). This means that the actors involved in a restoration project (including funding agencies) allow surprises and thus potential failure but have installed a structure that allows learning through these surprising events—such as focus groups, discussions, workshops, and repeated public meetings to assess the stakeholders' different views of the natural world. When openness is high and the acknowledged nonknowledge is taken into further planning steps, the robustness of experimental practice is also more likely to be socially and scientifically robust (quadrant 1). When both are low (quadrant 4), then the amount of ignorance is not realized, the actors involved believe that enough knowledge is available to act, and a top-down strategy of planning (no openness to different voices, no surprises, no "green time" strategy) is used. At this point, these parts of ecological design projects have already lost their robustness. When openness to surprises is low and the amount of acknowledged nonknowledge in relation to the amount of available knowledge is high (quadrant 4), then a research process will soon loose its robustness and will move into quadrant 3, since the former knowledge about the unknown is becoming useless because it is not experimentally tested via the fostering of surprising events. In contrast, projects with little or no explicit acknowledgment of nonknowledge and a high openness to surprises (quadrant 1) have the potential to become more robust since repeated efforts to learn (openness to surprises) can elevate the level of knowledge via the amount of nonknowledge becoming clearer and move into quadrant 2. However, since recent developments appear to have moved away from earlier experimental strategies, where the involvement of the public was rendered an important part of delivering a more robust legitimization basis, I added an arrow from quadrant 2 to quadrant 4 with a question mark.

In the early 1990s (and also prior to this date) until the eruption of the so-called Chicago controversy, the willingness to acknowledge nonknowledge was low, since the Chicago Park District thought it knew enough about people's attitudes toward restoration *and* the ecosystems that were to be restored. Furthermore, the openness and preparedness to surprises was also low (quadrant 3), since the surprise of public protests brought the restoration plans and activities to a halt for almost a year. After this phase, however, a new approach started. With a new preparedness to surprises, more nonknowledge (in the sense of unknowns that were specified enough to be considered for further action) was realized. Hence, the robustness of the process got stronger. Since the early years of the new millennium, the process appears to be robust enough to continue without ruptures or major breaks, since new nonknowledge was considered during each new step of the planning process. This also points to the aforementioned fact that a robust strategy is a means to building a tight connection between ecologists and the larger community to strengthen basic research *and* the human community's well-being.

The next chapter discusses a different type of case—the remediation and restoration of open mining pits in eastern Germany—although some of the material facts (such as the amount of soil moved or the sheer number of activities for redesigning a landscape) appear to point in a similar direction. Whereas the discussion of Montrose Point focused on the integration of surprises as a general pattern, the mining case in the following chapter takes a closer look at how successful processes for shaping the natural world explicitly specify what is not known. The typological differences between the two cases are rooted in their political starting positions. After 1990, most of the eastern German brown coal industry came to a halt. The period from 1990 to 1995 was a time when seemingly anything was possible, and it allowed planning and activity that were almost limitless in their openness to surprises in spite of an enormous amount of nonknowledge. The most obvious landscape change has been the development of many abandoned open pits into lakes. In the major mining areas in eastern Germany, some forty-five new lakes with an overall surface of 150 square kilometers are currently being designed and will be implemented in the near future. This will lead to a completely new image of the former landscape. The openness in the design process has changed considerably since the turn of the century, however, so that robust restoration and landscape development strategies, as they have been observed in restoration projects in the Chicago area, have become scarce.

5   Postindustrial Landscapes as Laboratories of Change

The new always happens against the overwhelming odds of statistical laws and their probability, which for all practical, everyday purposes amounts to certainty; the new therefore always appears in the guise of a miracle.
—Hannah Arendt (1958, 176).

After the unexpected fall of the Berlin wall in November 1989, reunification of East and West Germany became official on October 3, 1990. This date marks the rapid end of two sovereign states with radically different systems as well as the end of the wider divisions of the cold war era. The changes that were brought by unification altered the East German political, economic, and social systems, but they also affected the development and relocation of industries and settlements, land use, and landscape development based on brown coal surface mining.[1] At that time, the former state of East Germany, the German Democratic Republic (GDR), was the largest producer of brown coal (lignite) in the world.[2] Its goal was to become as self-sufficient as possible and to minimize its dependence on other energy sources. Most brown coal is burned for electric power generation, and mines are usually close to utilization sites for both economic and safety reasons. Until 1989, some 300 million tons of coal per year were produced in East Germany.

Coal provides a significant amount of the world's energy supply. Where coal lies close to the earth's surface, mining activities have radically altered the landscape in ways that are unmatched by any other human activity. In open-pit mining areas, the landscape has literally been used up. Production of one ton of brown coal requires the removal of 6 to 7 cubic meters of water and 4 cubic meters of overburden (mining debris) materials. Since most East German coal seams are near the surface, coal was extracted by strip mining. Strip mining exposes the coal by an advancing open pit or strip. The process starts by dewatering the complete mining area down to the layers under the coal to be mined. As the coal is exposed and extracted, the overburden from the still covered coal fills the former pit, and the strip progresses. Open-pit coal mining in East Germany was centered in the central states of Saxony and neighboring Saxony-Anhalt. In Saxony south of the city of Leipzig and the eastern Lusatian (Lausitz) region close to the border to Poland, the landscape has been devastated.

The city of Leipzig, which has a rich tradition of music, culture, and trade fairs, was the birthplace of the "peaceful revolution" in 1989 that lead to the reunification of Germany, but its landscape has been altered tremendously. With the help of the largest earth-moving machines in the world (large bucket-wheel excavators that are over 200 meter in length and up to 100 meters in height), brown coal mining swallowed up entire land-scapes, villages, and neighborhoods around and even within the city limits of Leipzig. Massive alteration of surface and groundwater was one of the many ecological consequences of this mining, so that in many regions the water table dropped by over 30 meters.

In a way, this chapter on the areas south of Leipzig presents a contrast-ing case compared to the restoration of Montrose Point because since 1990 Leipzig's openness to surprising events and nonknowledge has been excep-tionally evident but declined since the late 1990s.[3]

Europe's Largest Landscape Construction Site

Brown coal mining in Germany can be traced back to at least the four-teenth century, and since the early nineteenth century it has been a cata-lyst for industrialization all over Germany. After 1945, Germany's division into two fundamentally different economic systems (a market economy and a planned economy) led to the evolution of two German lines of development, which also affected its brown coal industry. Unlike the Western part of Germany (Federal Republic of Germany), the German Democratic Republic increasingly based its energy generation on brown coal mining. In the late 1980s, the respective shares in primary energy generation was 7.5 percent in the western part and 63.7 percent in the eastern part (cf. Berkner 2000, 186). Brown coal was East Germany's only abundant raw material for energy and for export to Eastern Europe and the Soviet Union. It was the main fuel that it used to produce electricity and heating.

Although environmental protection was institutionalized in East Germany's constitution in 1968, the country lagged behind in the development and design of natural areas and was unable to develop alternative technologies and fuel-efficient machinery. As Dieter Rink (2002, 77) put it: "Instead of switching from an intensive use of raw materials, energy production relied on the increasingly extensive exploitation of lignite—the only natural resource that East Germany had in abundance." West Germany, in contrast, was integrated into the international economic system and aimed at "a reasonable energy producer mix" (Berkner 2000, 186).[4] With 25 percent of the global output of coal, the East German state became the world's biggest single brown coal producer until the fall of the Berlin wall. This development went hand in hand with its goal of becoming self-sufficient in energy generation. Brown coal was also used as a raw material for the chemical industry because the country lacked a strong enough currency to purchase oil on the world market. This had a devastating effect on both air quality and the landscape in the decades after World War II, and since the 1970s, extraction has taken precedence over land revitalization, reclamation, and renaturing of the landscape.[5] The open-pit mining in the areas south of the city of Leipzig and in Lusatia near the Polish border southeast of Berlin changed the landscape on a scale that is unique in world history.

Open-pit coal mining often generates areas where coal is stacked and where coal has significant sulfur content. Such coal heaps generate highly acidic metal-rich drainage when exposed to normal rainfall, and these liquids can cause severe environmental damage to receiving water bodies. In addition, the waste heaps are subject to slipping. Overall, environmental protection and the protection of human health have been secondary. Most environmental problems were based on sulfur dioxide and black dust from power plants or emissions from chloride plants (cf. Murdoch, Stottmeister, Kennedy, and Klapper 2002). Furthermore, due to increasing evidence of heavily contaminated sites near settled areas and to the well known deficiencies in landscape recultivation and land restoration, brown coal mining had already lost some of its acceptability in the years before the fall of the

Berlin wall (cf. Herwig 2001; Rink 2002, 83).[6] Within a few months of the changes in the fall of 1989, open-pit brown coal mining became generally unacceptable.

Because of the political changes in 1989 and 1990, brown coal mining also became nonprofitable at the international level. After 1990, most of the surface mines were closed since they did not operate efficiently and were exposed to world market prices. The region south of Leipzig experienced a breakdown in economic and social structure that was unprecedented in Central Europe's economic history in terms of both degree and pace. The chemical industry that was based on brown coal, for example, faced the closure of all thirteen briquette-manufacturing facilities and 90 percent of the open-pit mines. As a spokesperson of the Regional Planning Office in Leipzig observed, "the carbon-chemical industry survived the introduction of the Deutschmark on July 1, 1990, only by a few weeks" (Berkner 2000, 188). Due to the abruptness of the shutdown of most parts / of the mining industry, no provisional or preliminary planning was available to be built on. Nearly 40,000 employees lost their jobs almost over night. As in other regions of the former GDR, workers found much of their identity in their work, and with the loss of work, identities were shattered (cf. De Soto 2000).

After 1990, a landscape that had been shaped over centuries by successive adjustments from woodland to farming to the mining landscape of the 1980s gave way to a *Bergbaufolgelandschaft*—literally, a postmining landscape. Today the moonscapes of mining pits have been transformed into recreational landscapes, consumer commodities for the inhabitants of nearby towns, and natural areas via ecological restoration. These old industrial sites today are still being transformed (figure 5.1).

All over Europe, old strip-mining fields are being regenerated in many different ways, but in most cases, the first choice is to redesign the target areas into more natural areas, even though this landscape is substantially different from the landscape that was present before mining began. Since the early 1990s, many mining pits have slowly been flooded—some naturally by rising water tables and others artificially by pumping processing waters (*Sümpfungswasser*) out of active neighboring mining pits or water from adjacent rivers. Flooding these enormous open-pit mines is different from flooding construction sites, and without knowledge of acidification or possible groundwater contamination, it was in many respects a novel

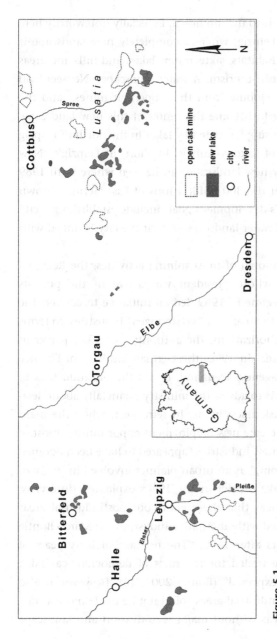

Figure 5.1
Open-pit mines in eastern Germany that are being turned into lakes (map courtesy of A. Bleicher)

practice and thus a venture into the unknown. Especially noteworthy here are the southern parts of Leipzig, where a completely new landscape is arising—including natural habitats, sixteen new lakes, and hills and areas for recreation and local green tourism. A water landscape—Neuseenland (New Lake District)—is developing from these open-pit mines,[7] and the City of Leipzig is planning to become the center of this new landscape. The new water landscape is supplemented by lakes in the north of Leipzig and in the adjacent state of Saxony-Anhalt. In short, in Leipzig's "New Lake District" a completely new landscape can be seen where until 1989 60 million tons per year of the 300 million tons of East German brown coal was extracted. Leipzig's development goals include establishing a city harbor as a gateway to the water landscape so that the city is linked with water tourism.

After 1990 and the shutdown of most mining activities, the Regional Administrative Authority, whose president was active in the protests against the East German regime in 1989, led an initiative to convert the southern areas of Leipzig into an area of "Westernized" industries. In terms of technical issues and authorizations, the authority was a key player in the development of this area. In 1990, the German Ministry of Finance appointed a privatization agency, Treuhand, to sell the old industries to new investors. Treuhand was to advise the ministry financially about decisions on rationality and risk assessments. In retrospect, when the main goal was to rescue old and create new employment opportunities, most of these attempts to establish new industries "appeared to have been doomed to failure and thus fizzled out," as an urban planner involved in many of the projects of the new lake district in the 1990s explained (interview, March 2006). It became clear that these enormous postindustrial areas could not simply be resettled with alternative industries. As Marco Bontje summarized the early years after 1990: "The heritage of forty years of socialist rule could not be traded for the rules of democratic-capitalist development as easily as expected" (Bontje 2004, 14). However, it also became clear that the old industrial areas could not be made more attractive for new residential areas without major recultivation and landscape design strategies.

Although the first years of the development of the southern parts of Leipzig after the fall of the Berlin Wall have often been called a failure, the region's openness to new possibilities and thus surprises helped it handle these severe problems of finding alternative industries. In 1994, the preliminary arrangements for EXPO 2000, the World Exhibition in Hannover, offered a proposal process for new landscape design projects and the possibility for external funding, which helped realize the potential for establishing a tourist industry instead of new manufacturing industries. In about 1993 and 1994, the idea formed that the former industrial areas south of Leipzig and its enormous open-pit mines could be turned into an attractive lake district. This project today is the largest landscape construction site (*Landschaftsbaustelle*) in Europe.[8]

I now focus mainly on the open-pit mines that are south of the city of Leipzig, especially the brown coal combines of Espenhain and Cospuden, which have been turned into lakes (see figure 5.2 below). The people in these areas found especially interesting methods of coping with unexpected events. First, this area was one of the most polluted in Europe. The village of Möbius south of the open-pit mines of Espenhain was known in East Germany as "the dirtiest place in Europe." Espenhain turned brown coal into chemical raw materials, and it sent huge amounts of dust, carbon,

Figure 5.2
Leipzig and its New Lake District (map courtesy of A. Bleicher)

sulfur dioxide, and other poisonous materials into the houses, clothes, and lungs of people. In addition, among the many villages and small towns that were destroyed and whose human population was resettled to different regions, the surface mining pits of Cospuden and Espenhain are unique since they are located close to the city of Leipzig. Cospuden is actually within the city limits of Leipzig. Their renaturing and remediation began sooner than that of the other mining pits.

The Design of a Lake District: New Nature in Postmining Landscapes

In 2005, Leipzig's former mayor Wolfgang Tiefensee, now the German minister of transportation, building, and urban affairs, announced that "Leipzig is a laboratory of change" (Tiefensee 2005, 6). This statement, which I am going to link to Robert Park's characterization of the city of Chicago as a laboratory (cf. chapter 2 above), refers to the numerous unique rearrangements and regenerations that the city of Leipzig continues to experience. As is discussed in chapter 2, a social laboratory is also a place for learning that is based on earlier surprising events. Taking our understanding of the notion of experiment and laboratory from the preceding chapters, Tiefensee's catchphrase can be rendered an apt description of the changes that have taken place in and around the city of Leipzig since the early 1990s. It also points to the connectedness between the working arena of scientists, engineers, and society at large.

The enormous changes that have taken place in the area south of Leipzig are based on the large amount of coal available. The whole surface of the landscape south of Leipzig sits on brown coal. Layers that are 20 to 30 meters thick lie under populated areas like suburbs of Leipzig and the adjacent city of Markkleeberg. Almost 50 percent of the western half of Saxony is sitting on coal. Before 1989, more than 10 percent of the world's brown coal production took place in this area. This gigantic mining and industrial complex was staged by more than 30,000 miners and energy workers and nearly 13,000 employees from the chemical industry (chemical basics) and power generation. Old maps of the area bear the names of many villages that were bulldozed during socialist times to make way for strip mines. Air, water, and soil were heavily polluted, sixty-six villages and small towns were devastated, and roads, historical buildings, fields, and woods disappeared forever (table 5.1 and 5.2). Since the late 1940s, over

Table 5.1
Villages and town districts south of Leipzig that have been completely destroyed

Town or Village	Time	Measures
Gaschwitz	1964–1965	Partial resettlement
Zöbiker	1971–1972 and 1978–1979	People in parts of the town moved to different districts
	1972–1973	First wave of removal and resettlement to northwest Leipzig (Umsiedlung Mühlsteg)
	1986–1986	Complete resettlement and devastation of town
Prödel	1971–1972	Removal of citizens
	1972	Devastation
Cospuden	1973	Resettlement
	1974	Removal
	1975	Destruction of buildings including a baroque church
	1981	Devastation

Table 5.2
Summary of inhabitants resettled in southern Leipzig

Time	Total Number of People Resettled
Late 1940s to 1960	8,500
1961 to 1970	6,000
1971 to 1980	4,400
1981 to 1994	5,200
Total	24,100

24,000 people in the area south of Leipzig lost their homes due to excavation. Estimates of the number of people who left their houses early because they feared losing them in the future go up to 50,000 people.

During the era of state socialism, living conditions were damaged by land consumption (see above), the destruction of natural sites, transportation routes being cut off, and ecological harm in the form of contaminated air, water, and soil as well as noise pollution. As Sigrun Kabisch (2004, 87), who lived in the area for over thirty years, colorfully summarized: "Everyday life on the edge of an opencast mine involved witnessing landscapes being ravaged on an enormous scale. Huge pieces of equipment such as excavators, spreaders and conveyor bridges are used to dig up entire tracts of land and remove overburden in order to mine the lignite underneath." The main part of the mines' overburden is removed from the original outcrop site by using bucket-chain excavators, which mix up different geological materials. The overburden material is transported to the spoil site by conveyor bridges, from where it is dumped without separating the different materials. The overburden layers thus become mixed, and former aquifers (underground layers of water-bearing permeable rock or gravel from which groundwater can be usefully extracted) become heavily changed, sometimes even destroyed. These overburden dumps become complex hydrological environments that are characterized by a large number of dissolved constituents and a number of potentially reactive minerals, each contributing to the chemical evolution of drainage water. By the late 1980s, this type of mining had swallowed up some 470 square kilometers of the landscape, of which less than half had been made reusable again.

In 1990, many welcomed the end of mining. As one resident stated in an interview (June 2005): "In 1990, it did not matter what was suggested as an alternative for mining, as long as the mining, the pollution, and the destruction of the land was stopped." Not everyone felt this way, especially the former mining employees who lost their jobs. Sigrun Kabisch (personal communication, April 2005) has referred to this process as "a dilemma in reverse," since in the GDR working in the mines was prestigious but the environment was very dirty. After 1990, however, the air became clean, the environment promised soon to become beautiful, but many people had no jobs and had to enjoy the new landscape on welfare income. Either way, the closing of the mines created a new beginning in the hearts and minds of Leipziger residents.

The step-by-step implementation of a project called Regional Network Neuseenland brought into being a unique experimental field (table 5.3). So far, for the remaining 150,000 inhabitants in the southern parts of Leipzig (Leipzig city proper has 500,000 inhabitants), this situation has been an enormous economic challenge but also a unique chance to master ecological change through innovative landscape design and novel technologies. There was no or very little time for planning and knowledge generation for the restoration and renaturing of these mine sites before they closed down. As the head of the regional planning office put it most succinctly (interview, February 2006):

For what we do here in the field of landscape reclamation, there is no blueprint whatsoever in terms of a nostrum or something similar that has happened before (so that we only need to adjust and take the knowledge from a similar case). . . . In 1990, we simply stood here empty-handed. There were no concepts about what we were doing and what was going to happen.

Furthermore, there is no system for assessing beforehand whether renaturalization, restoration, and other design activities will be successful. Even official documents such as reports from the Lusatian and Central German Mining and Administration Association (LMBV) (LMBV 2003: 19),[9] state that for the "solution of pressing problems as regards the rise of water level we ventured into uncharted terrain [Neuland] and had to fill the letters of the administrative agreements with life to master new organizational and substantive challenges." However, the process of fast flooding was started with the goal of developing a self-regulating water balance and opening up new development potentials for an economically and ecologically sustainable combination of classical and technology-oriented industries—of tourism and leisure economy as well as science and research (cf. Linke and Schiffer 2002). In the early stages, the knowledge that there were no blue-

Table 5.3
Overview of new lakes near Leipzig. The development of Lake Cospuden and Lake Markkleeberg is the focus of this chapter.

Lake	Dimension	End of Flooding
Lake Bockwitz	170 ha	Finished
Lake Cospuden	436 ha	Finished
Lake Grosstolpen	28 ha	Finished
Lake Hainer and Haubitz	545 ha	Finished
Lake Haselbach	335 ha	Finished
Harthsee	65 ha	Finished
Lake Kahnsdorf	112 ha	Finished
Lake Kulkwitz	150 ha	Finished
Lake Luckae and Goitzsche	869 h	Intermediate level in 1999; finished in 2060 (estimates as of 2007)
Lake Markkleeberg	252 ha	Finished
Lake Perese	589 ha	2051 (estimates as of 2007)
Lake Störmthal	733 ha	2011 (estimates as of 2007)
Lake Werben	79 ha	Intermediate level in 1999; finished in 2045 (estimates as of 2008)
Lake Zwenkau	914 ha	2013 to 16 (estimates as of 2008)

prints or historical reference points led to a remarkable openness of the actors involved regarding new and in this sense surprising developments. As a representative of a state authority pointed out (interview, January 2006): "Facing uncertainties does not always mean that that we can wait for more expertise." Consequently, the pressure to act led to a focus on trying things out and reacting to the outcomes.

Early research in geotechnology was spurred by slope failures in waste heaps in the 1950s, and since the 1990s soil mechanics and the development of large-scale models in geohydrology have been important areas of science and research spurred by the redesign of mining landscapes. The research undertaken for the design of the New Lake District has led to surprising effects and even failures—what Berkner (2004, 218) calls "elementary events," such as floods and rotational slipping—and these events have been major catalysts for research activities and the reorganization of research strategies.

For the county (Landkreis) this "Leipziger Land" meant that an area of 430 square kilometers was to be redesigned to become a flourishing region. The development of new economic activities requires the preparation of workforces for these new tasks. In 1996, according to data available from the Lusatian and Central German Management Company for Mining (LMBV), 31 percent of the landscape was farmland, and 17 percent was woodland. In 2050, farmland will most likely be just 20 percent, and 43 percent of Leipzig's southern regions will be covered with woodland and new forests (cf. table 5.4).

This case also points to the limits of experimental practice when the political culture becomes increasingly averse to surprises and falls back to

Table 5.4

Open-pit development and remediation in southern Leipzig in 1956, 1996, and 2050

	1956	1996	2050
Mining area	51%	48%, including the closed and not yet recultivated mining areas	0%
Arable land	31%	26%	20%
Woodland	13%	17%	43%
Water surface	4%	4%	30%
Additional areas	1%	5%	7%

a more linear or precautionary type of planning that does not allow the limits of knowing to be laid open. Instead, ignorance becomes an indicator *not* to act and to wait until enough (whatever that might mean) secure knowledge is available. As we have seen, in the early 1990s, the situation of the postmining landscapes south of Leipzig represented a unique opportunity to master ecological and structural change with creative landscaping. After several suggestions, models were developed in a competition between landscape planners and architects, the suggestions were assessed and screened, and citizen hearings were held. The first overall framework for the New Lake District was developed in 1991 by the Regional Planning Department of West Saxony (Regionale Planungsstelle Westsachsen). Based on the general framework of the regional plan, individual plans have been created via the issuance of licenses and city council resolutions aimed at attracting new investors (table 5.5). All developments and technical implementations fall under German mining law, which regulates the recovery of natural resources, the possible renaturing processes, and the usage of the postmining landscape in open-pit mining sites. According to German mining law, LMBV is in charge of the project management and obliged to prevent public risks. LMBV also allows the design of new socially accepted postmining landscapes.

The original plans from 1991 regularly had to accommodate social as well as natural conditions. In the summer of 2000, Lake Cospuden (water surface about 430 hectares) was the first of four lakes that extend into the southern city limits of Leipzig to be completely flooded. Lake Markkleeberg (260 hectares) followed in 2006, Lake Störmthal (740 hectares) will probably follow around 2011. For Lake Zwenkau (920 hectares), current estimations range from 2014 to 2016 (cf. figure 5.1). South of this area, the remaining twelve lakes are in the process of being constructed.

Table 5.5
Planning levels in the state of Saxony

	Planning Level	Outcome
State	Saxonian land-use planning	Regional development plan
Region	Regional planning	Regional land-use plan
Community	Town and country planning	Land utilization and development plan

Lake Cospuden, the first of these lakes, was named after a village in the area that was resettled in 1978 and 1979 so that the brown coal mining excavators could continue their activities (figure 5.2). Today, the lake's southern and western shores with its natural areas are suitable for bird-watching. On the northeastern and northern shore, near the edge of Leipzig, leisure amenities for the city's inhabitants have been sited. On the southeastern part, there is a port in the district of Zöbigker, which has developed into a popular water-sport center. Today the lake attracts some 500,000 visitors per year.

How did this result come about? The lake was flooded early, and its technical success and popularity with tourists evolved from several phases of experimental practice that integrated natural science research with social goals in the context of application. This practice helped people cope with surprises. In the first years after the mining activities ended, the Cospuden mine partially filled with rising groundwater from a glacial terrace aquifer. Since the early days of active mining, the groundwater level had been lowered, and for more than fifty years it had been down to 80 meters below ground. In 1993, the last pump for drainage of the pits was switched off, and water naturally returned. From 1995 to 2000, the Cospuden open pit was more actively recultivated and rapidly flooded by processing water from other active mining areas in the region.[10]

Anticipated Acidification and Surprising Heavy Metals

Based on estimations from 1991, the original plan expected that the flooding of Lake Cospuden would be finished sometime between 2005 and 2015. At the beginning of 1992, after the final mining activities in Cospuden ended, the open pit filled with both groundwater and percolating water from the slopes. This was still a slow process. Consequently, possible ways to accelerate the process of flooding were explored. With natural ground-water alone, it would have taken some thirty years or more to fill the pit. In several meetings that were held with investors, town planners, and concerned citizen groups, it was soon decided that natural flooding should be supported with groundwater from lowering the water table in neighboring surface mines in Zwenkau and Profen. As all of the stakeholders and technicians that I interviewed confirmed, at the time this was perceived as the only sustainable and socially responsible solution—even though many unknowns were involved and failures could be expected.

Zwenkau was one of the few mines that were still active in the 1990s, but it shut down in 1999. Water from surface mines in Profen was used beginning in 1998.[11] From this point in time, the water level was expected to rise by 3 centimeters daily. Based on this calculation by lake scientists and on knowledge that was gained from rare cases of flooding of open mines in other areas, the new estimation in 1998, when fast flooding started, was that Lake Cospuden would arrive at its destined water level of 109 meters sometime in 2001. Even one of the LMBV's informational brochures, *LMBV Konkret*, stated that "flooding at such a speed has no model anywhere in the world." Consequently, no one knew for sure when and how the flooding would be successful. These calculations rested on a large amount of ignorance, especially regarding the continuous and sufficient availability of industrial water. In the early 1990s, the actors involved (including the state funders) recognized that nonknowledge was the norm and agreed to take an experimental approach so that surprises did not have to be judged as failures.

After the pits were abandoned in the early 1990s, the constant pumping to keep groundwater below the natural water table to prevent flooding ceased, and the water table returned. The reintroduction of this water is the initial step in most acid rock drainage situations (cf. Frodeman 2005). Besides the challenge of the acidification of rising groundwater, heavy metals can mobilize from spoil heaps, and slope stabilities can be endangered (cf. Schultze et al. 1999). In cases such as these, it is not possible to predict when and in what amount heavy metals will appear since too many parameters are not available or not yet known. Furthermore, even if maximal knowledge of the full ecosystem would be available, it does not mean that you do know everything about the system's single parts. For the actors involved, this means that you learn to live with the unexpected because it is normal for things to go wrong. Especially in restoration activities that deal with soil contamination, people have regularly spoken about the normalcy of what is not known. As a representative of an engineering office stated: "Sometimes there is no time left. Then we have to decide ad hoc. A strategy needs to be changed lightning fast. However, by now all the actors involved are used to the fact that in every abandoned industrial site, some lack of clarity remains. Not everything can be documented" (interview, May 2007). In the case of flooding a former industrial site, heavy metals are a likely possibility given the sites' history, so the actors involved are able to formulate a concrete question about what is not known (nonknowledge) to prepare resources for coping with an antici- pated surprise. As Jörg ter Vehn (2006, 19) notes: "The amount of acidified material that is released is always the great unknown [die große Unbekannte]." A representative of the agency that is responsible for contaminated sites in the neighboring state of Saxony-Anhalt explained the starting position for dealing with contaminants in soil and groundwater (interview, June 2007):

In such a complex situation, we cannot say exactly what type of measures we need to take action against these hazards. We just don't know. Engineering companies have conducted excellent research, and academic research institutes with their more technical approaches have also integrated their results to help evaluate different variants. However, we still cannot say what exactly is going on with the groundwa- ter. But we need to do something. Or don't we? What then is our scale for doing something? This [not knowing] does not mean that we can stop the process and tell everybody to wait.

In a first step, lake scientists must persuade themselves and other actors involved that what is not known indeed cannot be known, at least not in a reasonable amount of time. What is needed in such a situation where risk assessments are not possible or not yet available is a clear reference point about what is unknown, so that a strategic question toward what is not known can be specified. Between an acknowledgment of what is not known (nonknowledge) and a decision to act, however, more is needed than trusting relationships (see below) between actors. As all stakeholders who deal with contaminants emphasize, intuition, "instinct," or some type of presentiment is needed. Since more often than not, as engineers and researchers explain, you need to "follow your nose," many field scientists who work with contaminants begin their work by literally taking a smell of something.

When heavy metals are present, severe soil contamination has been caused by harmful emissions from former carbochemical plants and bri- quette factories. As the groundwater rises and neighboring pits are flooded, contaminants can be affected by groundwater circulation. Thus, it was known where heavy metals would be found, if they occurred. During the early years of flooding, the acidity of the water changed remarkably in the pit that was to become Lake Cospuden. As a geologist involved in the project early on put it: "In 1993, it came as a surprise to find different grades of acidity even on a small scale, ranging from pH 2.5 to 8.5" (Schreck 1998, 71). Others talked about "a number of spectacular singularities" (Strauch and Glässler 1998, 10) when referring to this phenomenon. It took a while before the lake scientists could explain this surprising observation. They learned that it lay in the different types of dumped materials (pyrite content, carbonate mineral content) and different kinds of waste (brown coal ash) that were in contact with the water. After six months, the evolving lake even became more acidic in the southern part, and after a year, the lake still exhibited heterogeneous acidity. In 1995, the water appeared to have been stirred enough so that the lake's acidity was more homogeneous, but the water nevertheless had become more acidic in general. Based on these observations and available knowledge, it became clear that the flooding of Lake Cospuden had to be accelerated to neutralize the ph-value and to avoid further acidification. In a way, this can be characterized as a strategy for jumping ahead in the face of many recognized and acknowledged unknowns. It was agreed that the problem of dealing with the complex composition of different sources of acidity could not be solved by more research and in situ action on a small scale but that it needed an overall strategy to stop the entry of oxygen into the lake sediments (cf. Schreck 1998; Schultze et al. 1998; Strauch and Glässler 1998).

Another problem that arose during flooding of the open mines was the lack of slope stability, so that fast flooding became about the only option, since fast flooding can also stabilize slopes. However, the water for flooding that had been taken from the neighboring open-pit mine at Zwenkau since 1995 unexpectedly turned out to be acidic so that the pH-value dropped drastically (Abel, Michael, Zartl, and Werner 2000). Thus, the actors lowered the amount of water introduced into the new lake from Zwenkau since the goal was to have a pH-neutral water quality. After 1998, the main part of water was taken from the active Profen mine, which was much less acidic, so that the water could be neutralized.

The fast flooding introduced heavy metals into the water, however. This was a surprise that no one could explain immediately after the incident since it was hoped that stabilized slopes would also stop the influx of unwanted materials. The actors thus became aware of their own nescience, but it took some time before this insight (ignorance) was translated into proper nonknowledge. Only after historical investigations did it become clear that the heavy metals had originated from the activities of uranium mines north of the Zwenkau mine, although it is still unclear to this date if this happened due to a lack of wastewater treatment or to purposeful disposal of sludge into the groundwater.[12] The actors knew that in reference to their concrete implementation strategy, not enough knowledge was available. However, with the detection of the origin of the heavy metals in the uranium mines, they were able to name the point of reference of what was unknown (nonknowledge). Nonknowledge about the exact treatment for the sedimentation of heavy metals could be generated since its source became known and thus could be used in a meaningful way.

Subsequently, problems with heavy metals, whatever their origin might have been, were solved faster than anticipated. Indeed, the heavy metals had been adsorbed to iron and aluminum hydroxide adsorption, which developed during the neutralization process. Today they lay buried in the lake's sediments, and as all of the experts that I talked to confirmed, it appears highly unlikely (but not impossible) that they will ever escape sediments. Because of this, no particular treatment of the heavy metals was needed, since the reaction of the heavy metals was understood based on newly gained knowledge in the field of chemistry and building on previous investigations from other postmining lakes (cf. Klapper and Schultze 1995). However, a lake scientist who spent most of the 1990s doing research on lake flooding notes that "all of this is research in the open field. You only know something for sure after the implementation—here, the neutralization and the final investigations later on" (interview, June 2007). As an intended side effect of the neutralization process, the water from the Profen mine helped to solve the problem of heavy metals flowing in the water (figure 5.3).

Thus, although the actors involved knew precisely what they did not know (nonknowledge), they had a long way to go to transform this nonknowledge into extended knowledge. Today Lake Cospuden is an oligotrophic lake with very low nutrient levels but a large number of fish. What appears to be an anticipated surprise for the lake scientists can nevertheless be communicated as a (in this case, positively evaluated) surprise by the local media and the nontechnical actors involved.

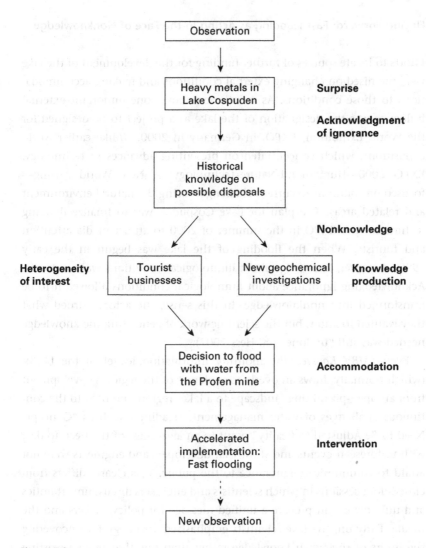

Figure 5.3
Unexpected heavy metals in Lake Cospuden and the opportunity for fast flooding. After accelerated flooding, the lake's acidification level fell, so that Lake Cospuden now has some of the best water quality in eastern Germany.

Fleeing Forward: Fast Flooding as Acting in the Face of Nonknowledge

Efforts to locate sources of further funding for the development of the lake were premised on changing external conditions and making accommodations to those conditions. As mentioned above, one important external influence was the designation of the lake as a project to be designed for the World Exhibition, EXPO, in Germany in 2000. Unlike earlier world expositions, which concentrated on presenting advances in technology, EXPO 2000—Humankind-Nature-Technology: A New World Arising—focused on solutions to current challenges facing the natural environment and related areas. The plan for Lake Cospuden was to finalize flooding in time for the EXPO in the summer of 2000 to attract media attention and tourists. When the flooding of the lake was begun in the early 1990s, however, knowledge about limnological conditions was miniscule. Acknowledging ignorance about limnological conditions allowed it to be transformed into nonknowledge. In this sense, the actors learned what they wanted to know, but the scientific work of generating the knowledge needed was still "undone" (cf. Hess 2007).

Even *LMBV Konkret*, the monthly information leaflet of the LMBV (which normally draws an overly rosy picture of the region's development from an open-pit mining landscape to a lakes region), pointed to the continuous challenges of water management. Headlines such as "Concepts Need to Be Adjusted to Reality" point to an awareness of the need to deal with unforeseen events and show that developers and engineers were not afraid to communicate ignorance to the public. This clearly differs from closed-door sessions in which scientists and engineers discuss uncertainties and unknowns and prepare a unified message for policy makers and the media. Early on, the lake scientists emphasized a strategy for uncovering the limits of their own knowledge rather than one that proved existing knowledge to be correct, and they clearly communicated their goals to the general public, as newspaper articles in the *Leipziger Volkszeitung*, the major daily newspaper of the region, show. In November 1993, for instance, one article claimed that in ten years, all Leipzigers would be able to take a bath in Lake Cospuden ("In zehn Jahren kann gebadet werden"), but the same article as well as other news clippings of the time pointed to major challenges and unknown parameters on the part of the lake scientists involved. Problems were generally communicated as opportunities.[13] Other articles stated that even provisional beach and landscape designs would foster a desire for recreational activities and that problems related to slope stabilities would not hinder the further development of the lakes. The general themes included scientists' uncertainties about the outcome of flooding and statements that more research would not help in the short term. Nevertheless, public support of the work of scientists and engineers in the context of application did not wane—quite the contrary. The limited knowledge and predictive capacities of science were not seen to be signs of poor science. Instead, the actors agreed on what was not known and took it into account for future planning—that is, they decided to act in spite of nonknowledge.

This attitude appears to be remarkable for several reasons. Although nondecision is impossible (since inactivity also implies a decision), knowing that surprising events will have consequences that actors and stakeholders might view as negative makes any decision difficult. In this case, however, making a decision in spite of nonknowledge was easier since the whole area fell under German mining law. The responsibility for dealing with possible failures and thus for being prepared for surprises lay in the hands of former mining companies, which here meant the LMBV as the successor of the privatization agency Treuhand. The role played by Treuhand was crucial since most decisions are made knowing that future consequences will be attributed to the decision makers (cf. Collingridge 1983; Collier and Lakoff 2008; Luhmann 1993). On this note, another representative of a planning department described a situation in which such a decision to act was made in spite of ignorance (interview, February 2006):

It might sound strange, but it was the only realistic possibility that we had. The other alternative would have been to try to move further along the usual path in water management, which would have been cost intensive with all the negative effects for regional development. Back then, we had to expect that things could go wrong. However, this can be seen as an opportunity.

Many actors involved refer to the years leading to 1997 and 1998 as a time of action and innovation simply because different possibilities were being tried out. At the time, it was known that the problems induced by flooding included acidification of the rising groundwater (cf. Berkner 2004; Schreck 1998; Schultze et al. 1999). However, in the early 1990s, an urgent need for action faced laborious hearings and the intricacies of the full planning repertoire. The main actors involved in 1992 were the district president, the City of Leipzig, the LMBV, the Regional Planning Department of West Saxony, and representatives of the communities and local investors, especially the soon-to-become recreational and tourist businesspersons. As all of the involved actors have noted, the only solution was to act quickly and knowingly in spite of many clearly unknown factors. In a way, the scientists and engineers involved sometimes appeared as if they were proud of being able to improvise in unexpected situations. As a representative of a regional planning office further exemplifies (interview, February 2006):

We knew that after four to even six weeks the reclaimer[14] would arrive and we would have to start acting. Otherwise, the chance for the design would have passed. The only solution was to get as many experts and stakeholders together at one table to assess how things *could* work and then raise one important question: Do we want to try it right now in spite of the uncertainties? That's what we did, and it worked fine.

Acting in the face of nonknowledge thus appeared to be a feasible way to move forward, especially if the actors wanted the lake to be part of the EXPO. As the quote above suggests, the unknowns were specific enough (nonknowledge) that an exhaustive set of future states could be specified, although the probability for any particular outcome was unknown. Acting in the face of such a specified set of unknowns, however, requires (as all lake scientists and engineers involved have confirmed) the right dose of intuition and nerve. Specifying what was not known helped them to be confident and move on. As a representative of the Lower Soil Protection Agency summarized the linkage between not knowing and acting (interview, May 2007), "when dealing with contaminants, you need a lot of courage, you need to be brave." William Miller (2000) argues that courage is generally based on doing instead of judging and evaluating.[15] However, based on the lake case above, it seems that intuition and courage (not carelessness) can be evoked if actors agree on a clear definition of and agreement on what is not known.

In this spirit, the activities at Lake Cospuden continued. Because the flux of water for lake flooding could be accelerated using industrial waters from a neighboring active mine, flooding was expected to be completed early in the summer of 2000, which meant that the lake would be finished in time for the EXPO. For some actors, this possibility, fostered by the advertisement for the EXPO projects, was a welcome surprise; for others, it was less welcome. The shop owners and small business owners who had built their dreams and future on tourism applauded the prospect. Many were former miners who were unemployed at the time, and some had been so since mining operations had stopped. They would be able to open their shops one tourist season earlier than expected (Linke and Schiffer 2002). However, some of the engineering and construction companies that were building the lake shores, foundations, and dykes and renaturing the natural habitats greeted the condensed timetable with some disfavor since they had to speed up their work considerably.

Trust and Nonknowledge: Research in the Context of Its Application

As the preceding section indicates, coping with issues of ignorance requires trust among key stakeholders because when decisions need to be made quickly, stakeholders needed to rely on each other's expertise and flexibility. As is discussed in chapter 2, trust can be seen a hypothesis between knowledge and nonknowledge (Simmel 1992, 393). Preliminary knowledge combined with nonknowledge can be a starting point for new planning activities. For Simmel, there cannot be any nonknowledge without knowledge. At least a minimal amount of knowledge is necessary before nonknowledge is perceivable. Nonknowledge is thus a prerequisite for trust since full knowledge does not require any trust between actors.

By 1998, decisions about the lake region still had to be made based on a considerable amount of nonknowledge. As we have seen, fast-flooding (via a pipe from the open-pit mine in Zwenkau and later from Profen) was undertaken based on the stakeholders' agreement that unexpected events were likely. However, this type of unknown is not nescience, which indicates a situation where the probability of consequences are not known and where a surprise may exceed any type of anticipation. It also differs from general knowledge about the limits of knowing in a certain area, which is inherent in the definition of *ignorance* that is given above. Instead, the actors involved in the design of the new lake agreed on what was not known and acted in spite of well-defined nonknowledge. Acting with nonknowledge, however, must not be confused with carelessness. The sharing of practical nonknowledge requires a careful and finely tuned consideration process among many stakeholders, across levels of governance, and (in the sense of a mode 2 type of research) across the science-practice divide. Overall, the negotiations among actors during the flooding process have been characterized by flexible decision-making processes and adjustments to cope with surprising events.

A further prerequisite for a successful implementation is dealing with the change of permissions (administration), flexibility in the change of plans (engineering companies), and a certain flexibility to redeploy capital for remediation (investors or public funds). These are crucial issues. If the discourse among the actors on a surprising and often potentially disrupting event distributes the accountability relations too much, a major consequence can be accusations and blame shifting, which can harm mutual trust. Those actors who attempt to make claims about what is unknown or what cannot be known can face a major problem because they can only point to the process of gaining more or extended knowledge in spite of unknowns—that is, by trying things out. This then can mean that actors involved will not be willing to stand up to their responsibilities and may treat surprises as unwelcome.

One is tempted to speculate that too much transparency about what is unknown can have a reverse effect because as the limits of certified knowledge become apparent, this information may weaken the trust in scientists. Indeed, lack of trust can easily bring a process to a standstill. Trust can be seen as a prerequisite to dealing with unknowns since without trust the stakeholders' willingness to consider surprise as opportunity for learning rather than failure, blame, and finger pointing is limited. When surprising events are expected, the responsibilities of the key actors (such as lake scientists, state agencies, and engineers) should not be changed or changed very minimally. This can be a difficult issue since flexible shifts in responsibility also belong to a successful strategy at coping with surprises. For instance, scientists sometimes need to act on behalf of policymakers, and the hydrologists working for the engineering office need to make decisions on their interpretation of data that were collected by academically trained scientists. Under such circumstances, tensions can easily arise among stakeholders, even when all are committed to reaching a common goal and by so doing to accepting surprising events.

The task is not easy, and there are no simple answers. However, considering the discussion so far, a robust strategy for identifying the limits and opportunities of responsibility distribution depends on an experimental process of carefully trying things out by fostering and integrating surprise and ignorance. Under circumstances of nonknowledge, no decision can be known to be absolutely right or wrong, so stakeholders need to be prepared for surprises so that they can afford to take opportunities and not follow a cautious approach—what Collingridge (1983) once called *hedging*. For Collingridge, hedging is a strategy for minimizing damage when the worst possible outcome is about to occur. It means formulating strict rules and eliminating all unknowns beforehand. At the other end of the spectrum, Collingridge saw *flexing*—creating flexibility to maneuver more freely and cope with ignorance. In other words, making small and quick decisions at the time is almost certainly better than dealing with the aftermath of catastrophic decisions later. In the lake case, both sides need to be taken into account: strict rules are needed to allow the actors to move freely and flexibly.

New informational devices can help. In the case of Lake Cospuden, a new information system was used experimentally under constantly changing geological, technical, and socioeconomic conditions. A newly developed geoinformation system (GIS) supported engineering, commercial planning, ecological possibilities, and approval preparation (Fritz and Benthaus 2000). For this strategy to succeed, all stakeholders had to trust that demonstrating their own ignorance would not have a negative effect on their credibility and the climate of cooperation. This is remarkable since the literature on trust points to the importance of preexisting relations that are a prerequisite for increasing the overall level of trust (e.g., Giddens 1990; Dodgson 1993), which would allow for such a type of handling the unknown. Although many of the actors involved had never met before the New Lake District project began, their relationships were built on trust.[16] Perhaps the stakeholders trusted each other because they did not have much of an alternative. The pressure to be successful and an everything-is-possible attitude (at least during the 1990s) was prevalent among the stakeholders and certainly helped their collaboration efforts. Sometimes trust in dealing with unknowns can become a reality not by choice but by the need to unite to create a sustainable future for a region. When the actors revealed their ignorance, they greatly benefited the necessary flexibility of the process. They also benefited the direct flow of information, which is a precondition to dealing successfully with ignorance. The aim thus was not to overcome or control the ignorance after more research was accomplished but to act and flood the lake in spite of ignorance and, by so doing, advance scientific research in the context of its application. This procedure did not undermine any type of "scientific optimism" or lead to growing cynicism and pessimism about research, as scholars in some fields of science studies have proposed (Felt et al. 2007; Gannon 2008; Wehling 2006). Taking the research involved at Lake Cospuden as a touchstone, laying open the limits of knowing for the scientists involved meant that they would not play the traditional know-it-all role but would conduct the research needed to move forward.

Overall, fast flooding is necessary for security reasons to ensure the stability of escarpments and slopes. When flooding proceeds slowly, escarpments need to be redesigned and rehabilitated—a much more complex and cost-intensive practice than flooding at a speed of 2 meters minimum per year. It turned out that fast flooding has a positive effect on both slope stability and water quality. During the design of Lake Cospuden, the water was not stopped, although for technical reasons it had to be slowed down for a few months at one point in 1999 (Berkner 2001, 53). This slowing down had to do with the groundwater pumped into Cospuden from the active mining operations in Zwenkau (see above). Some scientists warned the LMBV to pump less water. However, as a lake scientist involved points out (interview, June 2007), the open question was "when exactly the neutralization of the water's ph-value could be reached and, even more drastic, could it be reached at all." Examples from other, much smaller mining lakes had shown that local landslides and the loosening of sediments mobilized so much acidity that neutralization processes had to be postponed, which would have meant that the goal to finish the lake in time for the EXPO would not have been met. Even today, mine water from neighboring pits needs to be pumped into Lake Cospuden to counteract acidic influxes from former embankments of the mining pit.

In general, slopes at former open-pit mines are very steep, and responsible engineering companies cannot guarantee the long-term stability of flooded pits. Due to the sandy and liquefying dumped materials, the grain structure of the slopes often dissolved in former flooding processes. In some cases in the Lusatian region near Poland, slides and erosion happened in the form of retrogressive collapses, which developed backward as staggered fracture for up to 300 meters into the hinterland.[17] At Lake Cospuden, a sufficiently compacted embankment was needed. Given the time pressure, several research institutes in cooperation with engineering companies and the LMBV developed novel soil-mechanical models and new technologies of compaction by blasting. The in situ development of these technologies and their application was an enormous challenge given that the compaction process has to be carried out in several increments in accordance with the rise of the water (Fritz and Benthaus 2000, 262). On one side, the public demanded that the pit be flooded as quickly as possible, and on the other side, the scientists and technicians had to operate under heavy application pressure. The technical implementation had to take place based on ever-new ignorance since new situations opened up horizons of unknowns. In this case, it was knowledge about the limits of knowing (ignorance) how to exactly fathom slope stability and stability of embankments during superfast flooding because of a rise of water more than 3 centimeters daily. Some actors jokingly baptized this process as "flash-flooding."[18]

Overall, the speed of flooding forced engineers and scientists to learn and extend their knowledge about what was not known into nonknowledge. This fast type of flooding led to the development of new technologies for the reclamation of devastated landscapes (such as the air-gun compaction method) (Fritz and Benthaus 2000, 264), which are now used in similar cases of landscape design in former strip-mining areas in Eastern Europe and South America (Berkner 2004, 220). This shows that by experimenting with new technologies in local niche practices, after only a few years technologies can become a global standard and thus help to create "a collective knowledge reservoir" (Geels and Deuten 2006, 273). Furthermore, the pressure to accommodate changing social conditions, such as the EXPO, also led to new knowledge and expertise that were valuable to science. In the case of Cospuden, the consequences of flooding fostered knowledge generation on causal relations and on the proportions of iron, aluminum, and magnesium in the water, which was critical for flooding an opencast pit at this speed (cf. Berkner 2004; Fritz and Benthaus 2002; Tischew and Kirmer 2007). The new and extended knowledge can again lead to the perception of further ignorance.

In general, flooding is not a process easily planned. If actors want to refrain from doing nothing, they need to agree to act in spite of ignorance. In lakes that are not flooded with industrial water from active mining pits, a dry year (or a very wet season) can mean that flooding may be completed many years later (or earlier) than planned. This type of indefiniteness holds true for many of the other lakes that are scheduled to be flooded in the future. It cannot be predicted with any reasonable degree of certainty if some of the lakes in the region will be finished by 2015 or by 2025 or if the lakes whose prospective final water level is estimated to be reached in the second half of the twenty-first century will meet their goal.

Up to this point in time, the expectations of the shop owners, tourist businesses, and the LMBV to establish a fast and cost-effective rehabilitation of decommissioned brown coal mining and coal-upgrading facilities have not been disappointed. After all, this is a critical prerequisite for successful future utilization of these sites, especially to foster tourist activities. Thus, we can speak of a process that has become socially and scientifically robust by establishing experimental practices that are open to surprises as a strategy for acknowledging ignorance as a prerequisite for acting with new knowledge. The process can be called robust because it opens a way to produce new scientific knowledge, albeit firmly built on a socially accepted strategy.

Although plans had to be revised to accommodate changes in natural and social conditions, in the 1990s surprising turns were coped with successfully. The years 2002 and 2003 are examples that illustrate the natural boundary conditions of this type of public experiment. The year 2002 was the wettest in German history since the beginning of recorded rainfall, including the so-called flood of the century near the rivers Elbe and Mulde in eastern Germany.[19] One year later, in 2003, Germany experienced the driest and hottest summer since the beginning of taking records. Although in one year, the amount of water was higher than anyone needed or could have technically handled, the next year not a single drop could legally be taken out of adjacent rivers. After the availability of industrial waters from active open-pit mines stopped, river water became an important source for flooding some of the remaining lakes. If unexpected occurrences like those that happened in 2002 and 2003 are considered normal, then even the best scenarios and computer models are useless. Everybody knows this, but not everybody will admit it. The even stronger flooding of the Elbe River and parts of the Mulde River in the spring of 2006, which in some areas even exceeded the 2002 flood of the century, certainly is a case in point.

Despite the completion of the flooding of Lake Cospuden in 2000, stable hydrogeological conditions have not yet been established in the surrounding of the lake. As of the summer of 2009, the challenge of speeding up the implementation plan has been coped with, but now the lake needs to be accepted as a place for leisure and recreation in the future. Plans to support tourism include new accommodation facilities, the expansion of canoe rentals, and increased access to fishing and other recreational activities. The lake festivals and restaurants located along lakes that do not yet exist furthermore indicate the power that even potential positive change in the landscape can have for people.

In general, the actors involved took an experimental approach to landscape design and development. This entailed collectively agreeing that surprising findings in the design processes would be addressed as they arose and that if they were significant they would trigger additional debate and a new consideration of public views toward the project plan. Planning and implementation were understood as learning processes that were triggered by the discovery of surprising events. Another lake scientist summarized, not without irony: "This flooding process has been very instructive." We can thus speak of a process that has become robust by establishing recursive practices that were open to surprises and the acknowledgment of ignorance but that allowed new knowledge to be generated for further action—as well as new spaces for exploring the boundaries of new nonknowledge.

Another important issue for successfully coordinating ecological design is the institutionalization of contacts and information exchange. Although this might seem obvious, in many projects that deal with known unknowns, it is not the norm to see regular consultations of all involved institutions and actors to exchange information, to discuss new developments, and to agree quickly on new strategies in accommodation to a new situation. The official rhetoric is that science will deliver results based on certainty. This also means that all actors must communicate their own ignorance—not as failure but as a normal way of dealing with the unknown. Acknowledging the unknown in an experimental approach should be perceived not as failure but as opportunity to learn.

Further into the Unknown: Rising Water, Shrinking Population

In Europe, a major surprise that only recently has begun to be communicated is population shrinkage. This is affecting the sale of building sites around Lake Markkleeberg, which is much bigger than Lake Cospuden and was flooded after Lake Cospuden (which was fully flooded in the spring of 2006). The sales of attractive building sites close to what is now a beautiful beach at Lake Markkleeberg have slowed considerably compared to the sales of new houses and shops near Lake Cospuden. The reasons are simple—first, not enough people and a far second, the economic bust. Debates about the purpose of new holiday homes near Lake Markkleeberg have been heated.[20]

Because of the population decline, the flooding of Lake Markkleeberg, which is only a stone's throw away from Lake Cospuden, has become a major issue in the local and national media. Throughout formerly socialist East Germany, lower birth rates and outmigration often go hand in hand with ongoing economic decline. The east of Germany has lost more than 10 percent of its population since 1990 due to low birth rates and movement of labor to the west. Since the fall of the Berlin wall in 1990, more than 1.5 million people have left eastern Germany. The Federal Statistical Office of Germany has predicted that in 2020 some 13 million people out of the current 18 million will be left in the area of the former German Democratic Republic. Although during the last decade, the regional birth rate slowly increased and today has reached the level of western Germany, outmigration because of unavailable job opportunities is still an ongoing process.[21] Without enough people living near these new leisure and tourist attractions, no sustainable and lasting economy based on local tourism can be maintained.[22] As the mayor of a small town from the vicinity of Leipzig observed at a meeting to finalize the adoption of a new mission statement for the future of the region: "We are longing for and planning with tourists that simply do not exist."[23] Even conservative estimates state that almost 20 percent decline in the population is expected through 2020 even in areas as attractive as the New Lake District (Berkner 2004, 223). Furthermore, Saxony is Germany's oldest state in terms of its inhabitants' age. An aging population and a high unemployment rate are not proper conditions for the traditional tourist industry that focuses on sports and leisure facilities for ages under forty.

Close cooperation among ecological scientists, the wider public, local business owners, and local politicians is called for. Some actors, especially the new shop owners and operators of tourist businesses, have a long record of societal learning after the decline of the brown coal mining industry, and they are not willing to give up their dream of becoming a new tourist attraction. However, the challenges are enormous. Traditionally, regional planners and landscape developers expected some kind of population growth, and interventions are based on massive investments. In regions with population shrinkage, there is little investment, either from local businesses or from international firms. In times of population shrinkage, the relationships among population development, job opportunities, and land-use designation are interrupted and cannot be interlinked in the traditional way. In short, planning for population shrinkage is based on dwindling resources and not on massive investments.

Oswalt, Overmeyer, and Prigge (2002, 59) illustrate this with a simple example of a motor boat and a sailboat: Whatever the current weather conditions are, a motor boat can move through all kinds of water with relative ease because it uses an external energy input (an investment in motor fuel).[24] A sailboat moves with natural resources (wind) and does not need an investment in an external energy input. Controlling and steering the sailboat need to be closely linked to wind and weather conditions, which can change rapidly and can becalm the boat. Only by making careful observation and adapting to wind conditions can a sailor reach a particular destination. Depending on the direction of the wind, the preferred destination might be changed, which in the case of landscape development in times of population shrinkage means that plans and goals need to be continually revised and stimulate the production of new knowledge.

This can be compared with the concept of *refactoring*, a term that is borrowed from software engineering to describe a process of clarifying and simplifying the design of an existing code without changing its behavior (cf. Fowler 1999). Refactoring is undertaken to improve the understandability of the computer code or change its structure and design to make it easier for future maintenance and usage. Adding new behavior to a program might be difficult because of a program's structure, so a developer might refactor it to make it easier to use and then possibly add a new behavior. In this sense, one can grasp the future development of the lake district as a process of experimental refactoring: no new additions or investment to current developments can be expected, and conditions are not understood well enough to make decisions on a firm basis but need to be made under ignorance. This process can be described as an experimental reorganization of given conditions (Gross 2008).

In light of the extreme decline of population in former industrial regions of east Germany, the creation of, as Jörg Dettmar (2004, 32) termed it, a "deserted postindustrial wilderness" is the challenge lying ahead. From an ecological perspective, this challenge might not be hard to meet since, as recent studies show (e.g., Kowarik and Körner 2005), wilderness areas need to be only minimally maintained to make them attractive for citizens— contrary to some nature-protection strategies that want to keep the public out of "nature." However, the legal situation of the areas in Leipzig's New Lake District makes the situation more complicated. The juridical process of preparing and redeveloping (*Flächenrückentwicklung*) land for wilderness purposes is a long and difficult process. In general, communities and cities are expected to make long-term plans for the development of local areas. However, existing land-use designations are rarely changed or canceled since changes or even small modifications of existing building laws normally mean a loss of economic value for the owners of the land. This loss has to be compensated for by the community according to the law for the liability for planning damage (*Planungsschadensrecht*). The rules that require compensation after changes in land allocations are a reaction to the conflict between the interests of the community in flexibility and the confidence of the owners in consistent and binding land-use plans. This duty to compensate landowners affects the actions taken by a community, and the consequences are that land-use plans, especially in natural areas that are not allocated for economic or leisure use, are being postponed (cf. Köck et al. 2006, 220–225).

In addition to these legal challenges to flexibility with surprising changes, a further unique complication needs to be noted. Urban sprawl in eastern German cities and especially in Leipzig is going hand in hand with urban shrinkage. Although population shrinkage and lack of economic growth should be expected to operate against urban sprawl, the decline in population has not been accompanied by a proportional reduction of households and buildings (cf. Couch, Karecha, Nuissl, and Rink 2005; Haase 2008; Nuissl and Rink 2005).

Furthermore, in the southern parts of Leipzig, many people from the inner city as well as rural areas moved into new housing projects in suburbia such as the New Lake District. Urban sprawl in the Leipzig area has been meant an intraregional redistribution from the inner city to suburbia and a decreasing population. Although on the average the population is still shrinking, the consumption of the surface for new buildings is still rising so that the development of successional natural areas is becoming even more difficult. As Nuissl and Rink (2005) show, without the development activities in suburbia that created a huge oversupply of buildings, building plots, and business parks, perforation of the inner city would have been less drastic, and urban sprawl would have been unlikely. Although recent authors (e.g., Chew 2008; Tremmel 2005; Haase 2008) argue that there are many positive implications of population shrinkage, positive effects on urban ecosystems and biodiversity have not yet been verified enough to make any final conclusions.

Despite the growth of forests and woodlands in the area south of Leipzig, one third of the area is still officially approved for economic development, such as small businesses in the tourist industry and other small-scale enterprises. The remaining shop owners and small businesses cannot easily be moved to change populated areas to natural areas, although many of the deserted lots between the modern office buildings resemble wilderness. The ecological research station in the neighboring town of Borna-Birkenhain (Ökologische Station Borna-Birkenhain), which is an important actor in the design of the lake district south of Leipzig, has proposed an alternative method for renaturing active usage areas and former brown fields. Its plan to build new natural habitats ("new nature") focuses on attracting investors who might have an interest in implementing new natural areas in the former mining areas. The ecological field station in Borna-Birkenhain thus shows that the public experiment of designing the New Lake District has the potential to initiate some natural scientific research and to stimulate a new model for designing the landscape.

Since many of the regions in former open-pit mining areas are large in scale and also suffer from human population decline, they have a great potential for renaturing. Because different soil layers were removed in the process of taking off the top layers of the earth's surface, some of the remaining layers consist of sand with varying proportions of loam. Surprisingly, this has been a good opportunity for renaturing since a poor substratum has turned out to be a good basis for primary succession and thus new natural (*naturnahe*, "nature-close") developments. The chances for nature development have generally been high since the soils are also free of fertilizers and other biocides. The remaining and returning native species are an important issue since today many of the endangered species of eastern Germany (such as the oligotrophic and mesotrophic kind in lakes as well as rare birds in former factory buildings) can survive only in former open-pit mining areas since they are very narrowly adapted species.

Even more critical than at Montrose Point in Chicago, the birds in the New Lake District do not merely prefer the unwanted nonnative bushes but can survive only in former industrial sites (which are areas that are not normally seen as natural). The birds include many open-landscape birds (*Offenlandarten*), and butterflies (such as *Hipparchia Semele* and *Melitaea Cinnxia*), grasshoppers, and dragonflies are also notable. The southwestern parts of Lake Cospuden and its shoreline and most parts of Lake Bockwitz and Lake Goitsche in the north of Leipzig also need to be mentioned. However, a return to a historical condition is never aimed for (cf. Altmoos 1999; Geissler-Strobel et al. 1998). When the purpose of the ecological design strategy is to renaturalize or restore a certain landscape, financial incentives usually are lacking. Rather, the central government and local authorities are encouraged to support this type of natural design since it is understood as a step to a rising attractiveness of the area that cannot be evaluated easily in economic terms. Future strategies need to be assessed regarding their experimental approaches to demographic change.

As the situation today suggests, a laborious and time-consuming stakeholder-participation process (including negotiations between ecological scientists and concerned citizens) probably lies ahead. Whatever the stakeholders involved end up deciding, the process has to be able to acknowledge changes and allow new knowledge to enter into the process. Given the demographic changes in eastern Germany, the project to design the landscape around the lakes as natural habitats and wilderness areas will probably allow their use by humans for recreation but not for traditional local tourist attractions such as playing soccer and renting motorboats. This appears to be likely since Lake Markkleeberg, next to Lake Cospuden (see figure 5.2), was completely flooded in the spring of 2006.

Similar to the situation at Lake Cospuden in the 1990s, the decision to start to flood Lake Markkleeberg was based on a large amount of nonknowledge. In 1998, experts estimated that the flooding of Lake Markkleeberg would end between 2004 and 2013. No one knew when the lake would be finished, and so no one could predict when it could open for leisure activities, for ecological restoration of plant and animal species, or for attracting new investors for the land that was contracted out as commercial zone and industrial areas. However, planning and implementing were done in an experimental fashion since the actors involved knew that they did not have the option of waiting for final knowledge that could be used as a basis for precise assessments of when the lake would be flooded. There were three options—doing nothing, waiting for proper scientific knowledge, and moving forward experimentally. The latter course was taken. In 2000, two years after flooding was begun, the experts revised their estimates and set 2008 as the expected date to completion. By the summer of 2005, they revised the date to early 2006. After several postponements, the official opening ceremonies and festivities took place in July 2006. The overall pattern of dealing with unexpected (albeit somewhat anticipated) events and with gaps in knowledge (Frickel 2008) at Lake Markkleeberg resembles the pattern of the flooding and design of Lake Cospuden—and probably many other lakes. Many obstacles lie ahead, and these are discussed further in the final section of this chapter.

The Challenge of Keeping Surprises Surprising

The original plans to redesign the Leipziger county region have been recursively refined to take into account changing realities. However, in the first decade of the twenty-first century, openness to surprises and an acknowledgment of nonknowledge have been changing. The design of the New Lake District began with enormous openness given Germany's unique political and environmental conditions. The beginning of the 1990s was a time of enthusiastic open planning due to new freedom and limited regulation, even though planning and implementation became more and more top down and less reflexive and open.

The typical example that many actors involved report to back up the accuracy of this observation is the licensing procedure for water regulations. The city of Leipzig oversees these procedures in connection with the development of the new lakes. This procedure helps to clarify the legal situations of the stakeholders, investors, and other participating groups. Although some fifty of these licensing procedures now exist for water regulations in the eastern part of Germany, fewer than ten of these were finalized as of spring 2009, and not one was finalized for the western part of Saxony, where the New Lake District is being designed. The office for plan-approval and licensing procedures (*Planfeststellungsbehörde*) believes that each time new scientific knowledge in the area of lake design is produced, it needs to be assessed for inclusion into the implementation. The process normally follows a predictable pattern. The agency that is responsible for the redevelopment of an area of the design of a lake files documents, which are examined for plausibility. This is followed by the claims of various stakeholders (including expert claims and counterclaims), so that the process takes at least a full year. By the time the procedure plan is resubmitted, subsequent claims are being made. "This way you never reach a result," as an urban planner summarized the issue (interview, March 2006). Another actor involved in the planning and monitoring of the flooding processes in the 1990s confirms that the decision-making processes back in the 1990s "would be impossible today." The early processes seem to have been more advanced than recent ones in that they did not claim that more research will simply reduce the scientific uncertainties that surround lake design.

Since the early 2000s, the focus has been more on uncertainties and on the idea that more science means more safety and less on initiating a public discussion about the limits of scientific inquiry and the normalcy of ignorance as a prerequisite to experimental approaches. "What is needed," a representative of a regional development and landscape planning office has said, "is an arrangement that secures two things. Firstly, what should be decided has to be decided right now. What cannot be decided at the current date can be moved to a later date" (interview, July 2006). Andreas Berkner (2004, 223) further clarifies: "Despite progress in redevelopment and new implementation technologies, we have to calculate the unexpected. . . . Thus, we can expect that the delayed effects of mining activities, such as processes of acidification or cases of old and inherited pollution, will make an appearance." To put this statement into the terminology developed above, decisions need to be made based on a clear consensus of what is known, what is unknown, and what needs to be assessed later. Without this, nonaction will be the result, and a stalemate situation will occur in which no resolution to move on smoothly is possible. Some authors have referred to such a situation as a "lock-in" (Grabher 1993; Hassink 2005; Wesselink et al. 2007), where the very success of technologies and strategies for new knowledge production are also the basis for the standstill. In the development of new landscapes, the actors involved have gradually maneuvered themselves into such a lock-in where only ever-increasing efforts to produce certainty can keep the system operational. In a similar way, the promising new directions that began in the early stages of designing Leipzig's New Lake District in the 1990s have become consolidated. Since the early 2000s and especially since about 2005, the focus in science communication clearly is aimed at obstacles and details in planning. The current challenges in developing the new lakes illustrate how an approach that originally acknowledged ignorance and offered space for negotiations to move on in the face unknowns has led to a false assumption that certainty via expert knowledge is the only possibility to move forward. This can be linked to what Sheila Jasanoff (2005) has labeled the civic epistemology of Germany, which historically appears to have depended on great trust in expertise and expert witnesses. In spite of early attempts to the contrary, the political environment has become increasingly incompatible with public experimentation strategies. I suggest that this lack of acknowledgment of nonknowledge has been a major source for some of the waning public confidence in the possibilities of the landscape design of former strip mines.

In contrast, consider the Chicago case in the preceding chapter. The restoration of Lincoln Park began with a very top-down strategy. It was strictly regulated until the early 1990s but today is very open and experimental. At the beginning of most developments, many promising and not so promising options are put forth, but more often than not, chance plays an important role in the direction that is taken by an ecological design. Ignorance does not only appear to be a normal part but perhaps even the most important and even driving force of many ecological restoration projects.

Although the thesis of lock-in for the most part has been applied to technical and industrial development, Joseph Huber (2004, 295) has observed that it "seems to be a rather normal, actually universal process in the life course of any system that is gaining stability, i.e. consolidating its system structures, be they the structures of personality and human biography, the structures of organisms and technologies, or the structures of civilization and society." Thus, for actors to break out of a lock-in, they need a new and clearly feasible perspective of change and learning. Sometimes innovative changes can no longer be considered, however, given current infrastructure or network externalities. In the field of technology development, lock-ins can resist improvements (Hughes 1987). A "lock-out" can be forced by an innovation that makes the network effects of the prevailing technology obsolete. More likely, it will be fostered by an unexpected outside influence.

This general pattern also holds true in the case of landscape development in former mining areas. When early successes attract further investment of new resources, the actors eventually lock in to suboptimal outcomes, and this has been a general problem of postsocialist transformation processes (cf. Hüser 2006, 19). As a result of these early lock-ins that are based on early successes, further flexibility and experimental practices are limited. Lock-ins are different in landscape design and technology development, and breakouts are based on a different set of prerequisites. The debates on how to unlock regional development from a path dependency have focused on concepts such as the learning region or the learning regional cluster (Grabher 1993; Hassink 2005). However, in research-based processes, integrating research into the development process is prerequisite for a dynamic system. Via increasing control of its environment, the research process increases its internal reliability as well as its extension. As has become clear in the case of Leipzig's New Lake District, the new directions that were promised in the early stages in the 1990s have been consolidated and have almost reached a lock-in stage. For many actors who are involved in the process, the time since 2005 resembles such a lock-in, a final stage in the development of the lakes, although many new questions and challenges (ranging from the natural sciences to economic and legal issues) lie ahead. A citizen of Markkleeberg pointed out that "today the focus on risks and uncertainties is much higher than the perspective on opportunities. This has slowed down the development here considerably." As a regional planner who is a geologist by training and was involved in the design of the lake further explains (interview, February 2006):

Planning has reached a stage where discussing problems has become more important than acting. . . . A planning office or a consultancy firm always gets the job when it can lay bare the problems most convincingly. This leads to a process of nonaction. Pointing out problems is legitimate and important. However, in recent years it has led to a stage of nonaction since it eclipses the ability to see alternatives and opportunities.

Unlike in the 1990s, since the early 2000s, many actors appear to be increasingly unhappy with economic, ecological, and aesthetic developments in the region. Since ever more certainty is required before investors or local businesses can take action, many planning and implementing procedures have come to a standstill (and in many areas, the brownfields are simply deserted). As most lake researchers involved today confirm, scientific uncertainty now compared to the early 1990s is much lower. This appears to be a known paradox. The more accepted knowledge there is, the more the communication about emerging risks and uncertainties can arise. It makes actors shy away from flexible decisions. The amount of ignorance—as in the example of Pascal's metaphor of knowledge as a growing ball (cf. chapter 2)—has obliterated and oppressed the actors. To use Simmel's description, the objective culture, although invented by human skill, has turned against the usage of knowledge for practical purposes. This tendency shows that successful coping with ignorance must always be coupled with openness to surprises and the ability to transform ignorance into nonknowledge.

In the view of many of the actors involved today, communication about these issues with the public and the other stakeholders has changed. In Leipzig, a false assumption that sufficient knowledge will soon be available leads actors to avoid accepting surprising events since the rhetoric of scientific safety and certainty does not allow accidents. Unfortunately, the public mistrust of science can sometimes be attributed to the failure of scientists to account properly for the domain of ignorance. In this vein, a regional planner and representative of the state of Saxony's Land Use Planning Department stated (interview, July 2006):

Today, compared to 1990, there is much more accepted knowledge about the shaping of postmining landscapes and the flooding of open-pit mines. Nevertheless, we are further away from absolute knowledge (which is impossible, anyway), and I am sure that in the near future (perhaps I should not say that I am looking forward to this time) we will face quite a few more challenges and surprises that will shape the development of our landscape.

Currently, people are not prepared to respond to unavoidable surprises. Figure 5.4 illustrates this decline in preparedness and in openness to surprises from the early 1990s to the present, using the same format as was used in figure 4.4 in chapter 4 where the case of Montrose Point was discussed. Figure 5.4 illustrates how a robust experimental process can decline and turn into a lock-in situation that needs a break out. Today, the lock-in situation in the southern parts of Leipzig has generated a kind of inertia, and the early freedom has been lost. It is an example of how treating

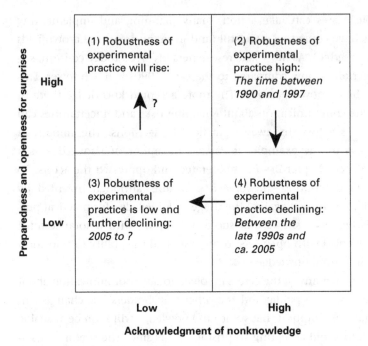

Figure 5.4
From experimental robustness to a top-down master plan of nonaction

surprises as anomalies and ignoring nonknowledge can create a lock-in process. What is needed might be a radical political or economic change that can lead to a lock-out. The figure shows how benefits of experimental strategies have decreased with the lowering of preparedness and lack of openness to surprises.

In retrospect, we can talk about three phases in the development of the lake district since the early 1990s—1990, 1997, and 2005. In the first phase, which followed the political turmoil of 1989 and 1990, the expectation of surprises and the acknowledgment of nonknowledge were very high. Indications of robustness were that residents opened windsurfing shops and other businesses because they were counting on the recreational and tourist business opportunities created by the new lake district. Scientists and engineers moved forward with the flooding process and slope stability, and both practices resulted in many new research results (such as in the development of the water quality in pit lakes).

Between 1997 and 2005, the amount of recognized nonknowledge was still high (quadrant 4 in figure 5.4), but there was no longer a mindful management that was open to surprising events. The amount of non-knowledge was increasingly bracketed out and not included in any meaningful way in discussions of future action. The scientists and engineers involved believed that enough knowledge was available to act, and when they felt that there was not enough knowledge, then no action was taken. The openness to surprises and the acknowledgment of nonknowledge were locked-in. A prior knowledge base and clear knowledge about what is not known (nonknowledge) have been good prerequisites for robust strategies, but the denial of surprises has led to a declining robustness.

Although openness and realized nonknowledge was high from 1990 to 1997 (quadrant 2), legal regulations, master action plans, and simply waiting for more knowledge have led to a decline of robustness (quadrant 4) since the mid- to late 1990s and have moved the process to quadrant 3 since the early 2000s, when many developments moved into different sometimes opposing directions. Today, the ecological design project for building a new lake district has lost its robustness. In the fall of 2007, however, there was a public discussion about a new mission statement and the future of the New Lake District in Leipzig. At the meeting, Leipzig's Regional Planning Office made an official statement calling for a less sur-prise-averse view (quite surprisingly for me, this explicitly was the termi-nology that was used during the discussions) and the will to act in spite of uncertainties. Thus, there are question marks in quadrant 1 and 3 in figure 5.4.

In this and the preceding chapters, the notion of robustness has been used in several contexts. Usually, *robustness* has referred to the quality of a research and design strategy, both in everyday life as well as in academic or scientific contexts. Robustness is needed to reach a goal even when surprises are the norm. Contrary to the notion of robustness in the field of engineering and cybernetics, robustness here is not treated as an abso-lute target but is able to be reified and accommodated in a recursive process. Robustness encompasses the stability and reliability of this itera-tive and recursive dynamic of practice, even if unanticipated occurrences lead to surprises, the original interests change, and societal factors inter-vene. As we have seen, robustness cannot be an indicator of knowledge per se but is a process that includes strategies of integrating different actor configurations, which are able to include and filter out the relevant interest constellations.

In successful projects of public experimentation, the wider public must become part of ecological implementation. Many commentators have stated that in knowledge societies citizens are called on to make decisions even on complex environmental issues (cf. Felt et al. 2007; Jasanoff 2005). Prior to Germany's reunification in 1990, policy-society relationships were dominated by a top-down approach in decision making. Although today public participation in Germany is encouraged and directly facilitated, the eastern German culture assigns a low value to public participation. A task for the next step in the new lake design project could be to turn concerned citizens into participating actors since the project in the southern parts of Leipzig (like the restoration at Montrose Point) continues to develop as it moves from one phase of the implementation to the next in response to changes in actors, policies, and natural factors. To a large extent, the local media informed people, but information was transferred in a one-way direction and was done after major decisions had already been made.[25] The integrity of the design process of the New Lake District south of Leipzig might become robust again after surprise and nonknowledge are skillfully used in planning and implementation based on the integration of all stakeholders and the implementation of strategies that turn concerned citizens into active agents.

In the medium term, usage and rehabilitation strategies in post–brown coal mining areas must transition from federal state support to the sponsorship of private investors, grant funding, community, and local authorities. For scientific research, stakeholder interests, and concerned citizen groups, resource bundling is necessary to cope with the new financial and sponsorship networks. After all, research in these new landscapes can be supported and built up only when research institutions deliver results that are relevant for real-world problems. This means that acting in spite of ignorance via the acknowledgment of nonknowledge will become ever more important in the future.

III Outlook

6 Welcome Surprises and New Edifices of Knowledge

The great risk of our time is being overtaken by inevitable surprises.
—Peter Schwartz (2003, 236)

Today, many people argue that science should remember its roots and original calling to be a servant of society and to focus on contributing to societal innovation and progress. When scientists communicate to the public, however, there remains the temptation of "trumpeting certainty and consensus, while refraining from clearly communicating what is not known" (Pielke 2005). Debates on the limits of knowledge are often avoided, presumably on the assumption that this would undermine stakeholders' trust in research. This, in turn, feeds the public's increasingly unrealistic expectations that science will deliver certainty and safety. When uncertain results are presented, science's former authority declines. This is unfortunate since recent research in science communication and public trust in organizations (e.g., Terwel, Harinck, Ellemers, and Daamen 2009) has shown that even unedifying results can increase trust in an organization as long as the arguments brought to the public are perceived to be honest and congruent with inferred organizational motives. One message of this study is that openly admitting and acknowledging that ignorance in initiatory practices such as ecological restoration and landscape design is part of "honest science" might foster new trust in science and its organizations. Trust can persuade concerned citizens and practitioners to participate in public scientific projects and to self-mobilize to initiate actions that are framed by science. The idea of knowledge production in public can thus be understood in the sense that John Dewey and also Robert Park used the term *public*. For Dewey (1988), the public is a group of people who share a common interest and apply common strategies and solutions to that interest. Scientific information thus does not simply go to various publics; it also creates them and in turn is created by them.

Practitioners in landscape design and ecological restoration have increasingly tended to pursue adaptive or flexible management strategies instead of strategies where the same original measure is applied in each decision period. However, under flexible management, a learning-by-doing approach developed or, in some cases, was moved back to a more linear model due to different restricting boundary conditions. The Chicago and Leipzig cases have shown what this can mean in the handling of ignorance and surprise. This study has established a working model and a description for an alternative route, informed by classical and contemporary social theory on handling surprises. Disruptive surprises have been framed so they can be fostered *and* controlled in one process via recursive networks in an experimental setting. In this final chapter, I further explore some of the social and theoretical value of considering public experimentalism as a fundamental and important part of social life. By doing so, I also point to the limits of experimental practices in dealing with ignorance and surprise by embedding and extending a discussion on the knowledge society with recent concepts of second or reflexive modernity.

Modernity and the Unanticipated Consequences of Progress

Modernity and the term *modern* refer to many different things. Charles Lemert (1993, 27) suggests defining *modernity* "as that culture in which people are promised a better life—one day. Until then, they are expected to tolerate contradictory lives in which the benefits of modernity are much greater than its losses." This definition also points to the core characteristics that Ulrich Beck attributes to "first modernity." Beck draws a pivotal distinction between first modernity and the contemporary world. Beck and his British colleague Anthony Giddens differentiate a first modernity from a second reflexive one, which correspond to a correct observation and criticism of the obsolescence of institutions for coping with the uncertainties presented by developments in science, ecology, and technology. Both Beck and Giddens understand the current time as an enlightenment of the classical modernity in the sense of a modernization of industrial society. For Beck, Bonss, and Lau (2003), postmodernists are interested in "deconstruction without reconstruction," whereas their proclaimed second modernity is about "deconstruction and reconstruction." In Beck's and Giddens's idea, modernity has not vanished but has begun to modernize its own foundations—that is, it has become reflexive. Beck claims that the notion of reflexive modernization disenchants modernity's own taken-for-granted premises and thus signifies a heightened awareness that mastery of the modern world is impossible. In this view, the modernity of the classics of sociology and related sciences was characterized by institutional rule: monolithic institutions set rules and guidelines, and the populace accepted and followed those rules. Furthermore, one of modernity's major characteristics—obsession with predictability—is diminishing as the aspects of the second modernity are being entrenched. According to Beck, modernity in the late twentieth century, as a precursor to second modernity, distinguished itself from former times by the disintegration of institutional rule making and the unwillingness of individuals to follow them. The second modernity has brought the immanent, unpredictable nature of potential technological and ecological uncertainties, such as the unintended side effects of nuclear energy, climate change, genetic engineering, or new chemical implementations.

Beck and Giddens observe that unpredictability, decreased control, and unintended side effects are the driving forces of contemporary societies. Giddens even describes *high modernity*, a term that he prefers over *postmodernity* or *second modernity*, as a "juggernaut" (Giddens 1990, 139, 151–154), a relentlessly driving and sometimes grim "engine" that can only partly be steered by human society. For him, the unpredictability and uncontrollability of the modern world have made it a "runaway world" (Giddens 2000).

Beck looks for a solution that calls for more opportunities for the public to engage in a reflexive modernization. Beck et al. (2003, 3) also claim that the notion of reflexive modernization does not signify an "increase of mastery and consciousness, but a heightened awareness that mastery is impossible." Reflexive modernization "disenchants and then dissolves its own taken-for-granted premises." Giddens (1990) calls for active trust relationships since "trust" will increasingly be the key to functioning relationships between the wider society and different expert systems. Central to Beck's and his coauthors' concept is the idea that first modernity is becoming more reflexive about itself as a conscious response to increasing risks and the unanticipated side effects that are generated by the processes of industrialization. By having reached the limits of modern developments, contemporary societies—at least those in the developed world (cf. Lee 2008)—are entering a new phase that should be called "second modernity." Unlike the earlier phase, new uncertainties and side-effects are specific to the risk society of second modernity. They are different from earlier hazards in the sense that they are virtually uncontainable (spatially, temporally, and socially) and are unable to be accounted for (Beck 1992). Today, Beck and others argue, modern society no longer takes for granted the foundations of modern growth, which since the second half of the twentieth century have been undermined. In this view, acknowledgment of the unintended consequences of modernity appears to be necessary for social change. What does this mean for our idea of public experimentation? As outlined above, our experimental approach needs to shift partial decision-making authority in the design of landscapes or solving environmental problems to engaged, local publics and to require a central state to pool information across experiments to help local experimenters understand the outcomes of other local experiments. This initial standard setting can be viewed as the moment at which citizens shape or frame the normative means of achieving public policy (such as targeting water-quality standards or environmental justice) (Elvers, Gross, and Heinrichs 2008). Deliberative standard setting should require participants to formulate their positions and their expectations with respect to on-the-ground experiments. This allows actors in future deliberative processes to demonstrate or question the coherence of these positions and provides a source of learning from surprise. If the world of objects and relations is a product not just of human interpretation of the world but of active interaction with it, then the procedural remedy for habit and routine is routine experimentation (Dorf and Sabel 1998). The argument is not that one true truth will be found via such experiments but that the process should generate conditions for greater societal learning to cope with unavoidable ignorance and surprise.

Research as Application: Toward an Experimental Knowledge Society

Attempts to determine the features of the emerging knowledge society the conception of an "experimental society" or "society as a laboratory" (Beck 1999; Krohn and Weyer 1994; Miller and O'Leary 1994) are reminiscent of some of the ideas of the Chicago school. The laboratory is perceived as a novel form of innovation where scientific research increasingly erases the received institutional boundaries between science and society. The only reliable avenue for determining the safety, usefulness, or effects of, for instance, a certain technology on the natural environment is to test it in the field. Because knowing all obstacles in advance is not possible while working within the world of the laboratory, scientists need to leave the lab, which is their accustomed sphere of exoneration from liability for the consequences of their activities.

The expression *knowledge society* should allude to the application and circularity of scientific knowledge and also to the production and recombination of all kinds of knowledge in new settings of knowledge work (Carrier, Howard, and Kourany 2008; Krohn 2001; Stehr 2001). If these observers are correct, then the experimenting society is already here, although it did not emerge from social science recommendations in the service of society and government. Rather, in an experimental knowledge society, conventions and norms are increasingly replaced by decisions based on expert knowledge and situation-specific experience. At the same time, knowledge producers and researchers operate in nonscientific environments, and if the settings of decision making are not locales of science in the traditional sense, they import and use methods of investigation and research (e.g., Herbold 1995; Levidow and Carr 2007; Lezaun and Millo 2004). Since new knowledge always creates new research questions, the application of knowledge is associated with the generation of new ignorance. Thus, ignorance is becoming one of the key indications for a knowledge society since processes of experimentation move back and forth between knowledge and ignorance.

Moving beyond the divides between experts and laypeople, many projects in ecological restoration and landscape design have opened up new ways of considering what a democratization of (ecological) science outside the laboratory could look like (cf. Light 2006; O'Brien and McIvor 2007; Moore 2006). Successful restoration projects need a public that is ready to acknowledge ignorance in all ecological practice since the unexpected is part of living in nature. People need to be ready to participate, to trust each other, and to take responsibility for the sometimes surprising side-effects of their actions. Otherwise, ecological restoration projects such as the introduction of the wolf in areas populated by humans will not work.

Considered in these terms, the return of wolves in eastern Germany and the conflicts over their return (discussed in chapter 2) is a real-world example that highlights the challenges that are inherent in many restoration projects. Furthermore, it shows how the openness to surprising events assumes a coincidental and sometimes disruptive nature of the course of things to happen. To transform the awareness of ignorance into nonknowledge illustrates how ecological design projects can become successful experiments that knit together academic research and the solution of real-world problems.

While acknowledging that the role played by science in landscape design and urban restoration projects remains critical, the case of Montrose Point in Chicago (chapter 4) illustrates that practitioners or stakeholders who have acquired their own types of "contributory expertise" (cf. Collins and Evans 2007) are able to account for successful projects and even new knowledge that is relevant for science. This phenomenon connects to debates on the democratization of knowledge and the public understanding of science as well as increasing social and political concerns about the values and interests of lay publics that are affected by scientific knowledge (cf. Brown 2009; Kleinman 2005; Moore 2008). Such a view, which sometimes is referred to as a shift from government to governance, draws many heterogeneous actors into the implementation and process of science in the making. Nevertheless, it cannot be based on the institutional conditions of unrestricted scientific knowledge production and thus has to address several challenges.

The practices of heterogeneous actors who are involved in landscape design typically have to be iterative or recursive, in that they alternate phases of knowledge production and knowledge application. *Iterative* here refers to a reflexive and cyclical practice that has a commitment to the continuous improvement of a process. *Recursive* refers to an iterative practice in which results are continuously disseminated among all involved actors of a design project so that the process can reach robustness both for scientific results and social acceptance. In this sense, *iteration* and *recursion* refer to the following process: the result of an implementation is taken as the starting point for further observations based on previous forms of knowledge production as well as negotiations over seemingly unscientific issues such as cultural values and aesthetics. In this context, *practices* can be most broadly understood as arrays of human activity that can also include work and play with material objects, even natural ones like trees or birds. For the historian of technology, Edward Constant, recursive practice is characterized by "alternate phases of selection and of corroboration by use. The result is strongly corroborated foundational knowledge: knowledge that is implicated in an immense number and variety of designs embodied in an even larger population of devices, artifacts, and practices, that is used recursively to produce new knowledge" (Constant 2000, 221). Besides the production of new knowledge, however, recursive practice in fields outside the strict realm of technology development and basic research in the context of this study have included the attainment of socially robust processes.

In general, after a sudden or unexpected event, ignorance becomes apparent. The right strategy cannot be doing nothing or waiting until secure knowledge is available. The participants have to start acting under increased attention from others to the consequences. Many previously agreed on steps need to be accommodated to changing or only little understood social and natural conditions. In an ideal process, the experimental practice needs to be continued until a solution for the social context is reached. The evaluation of the results of an activity refers to the previous assumptions and hypothesis. It is a circuit that feeds back some of the results of an implementation. Consequently, such recursive experimental practice has to include the acknowledgment of unknowns.

Although recursive flexibility in landscape development (in which measures and adjustments are selected in each decision period) has been increasingly perceived as superior to fixed management methods, this study examines the relevance of ignorance and surprise at the core of flexible, experimental processes. Instead of emphasizing certainty, the more promising path discussed in this study is an experimental approach that builds on deliberate steps of implementation and practical interventions. With an experimental approach, knowledge about what is unknown can be fed into each subsequent step of an implementation to expose it to further observation and to turn recognized nonknowledge into extended knowledge. The Chicago school of sociology's classical notion of "society as experiment" has been the touchstone for developing a model of public experimentation on how surprises and ignorance can be integrated. This might even include an expectation or even aspiration that surprises will occur, as was set out by Georg Simmel. By foregrounding the unexpected, a sociology of surprises implies that humans are neither entirely in control nor entirely at the mercy of society, technology, or the natural world. It reconstructs the unexpected as something that eludes modern society's mastery over its social, built, and natural environments through rational planning and science. Irregularities are not be treated as unwelcomed anomalies but need to be incorporated into social life and practice as a central issue and treated as entirely normal. This normalcy, however, must not develop into what Diane Vaughn (1996) has called the "normalization of deviance." Vaughn analyzed the development of a culture of incrementalism in the decision to launch the *Challenger* space shuttle in 1986. The fateful decision was based on poor judgment of anomalous events. Unlike disastrous decisions and mistakes such as NASA's *Challenger* failure, uncertain human activities in the natural world (such as ecological restoration) do not have to be fatal. This is certainly a major difference between some kinds of experimental approaches outside the laboratory. Even so, failures in ecological restoration activities can have aesthetic and economic consequences, as well as ecological consequences for biodiversity or the functionality of ecosystems in general. Although ecological restoration activities can be recognized as inherently uncertain, they can be modularized and organized in a way that allows surprises to become opportunities to learn rather than failures (Lee 1993, 56). Organizational theorists have long since proposed that when skillfully brought into action, an organization's own failure can provide important lessons and the experiences that are necessary for further development (e.g., Kim 1998; Sitkin 1995). This insight has also been taken up in publication strategies in the field ecological restoration. Richard Hobbs (2009, 2–3), editor of *Restoration Ecology*, argues that since "failure is often a necessary basis for subsequent success," research publications on restoration should include much more results that were "not an out-and-out success story."

Surprises are unavoidable, but what they may be and whether they are beneficial or harmful cannot be resolved by further research that is detached from the context of its application. Since sound knowledge by definition can be produced only in laboratory contexts, questions always remain about whether the knowledge produced under such artificially controlled conditions adequately represents the more complex, variable, and less controlled conditions of landscape design processes. Openness to surprises is a strategy that can be understood as the antidote to inertia and the belief that more research will lead to more certainty. As the cases in this book have shown, the contrary appears to be true: scientific knowledge is never completely settled, and it is always possible to invest in contradicting expertise that identifies conceptual shortcomings or open theoretical questions that serve to delay decisions. The application of knowledge is associated with the generation and management of new ignorance. At the same time, new knowledge allows us to see and better define new nonknowledge.

For Robert Park, experiments are undertaken by society. When he characterizes the city as an experiment, Park does not merely mean the experiments of city planners or social workers who take society as their object of study but rather the experimental character of societal development. Today's emerging knowledge society thus is a society of experimentation that is based on the handling of surprising events and the acknowledgment of ignorance typical for modern society. It builds its existence on certain kinds of experiments that are practiced outside the special domain of science. The notion of experiment postulated here is a deliberate intervention that is undertaken by a rapidly developing society that sets up institutional conditions of action without being able to control their dynamics or attribute events to human decision making. Unfortunately, the generally negative attitude toward ecological issues and science in general can easily obliterate many positive attempts of experimental strategies and their potential for environment issues (cf. Furedi 2006). After all, assessing the positive potentials of human societies' relations with the natural world also means having a tolerant attitude toward public experimentation and new forms of participation.

Understood this way, the experimental nature of modern society is transformed from an evolutionary process to an institutionalized strategy, which includes all kinds of political, cultural, and aesthetic components. The understanding of society as experiment thus broadens the notion of a knowledge society to one that is firmly anchored in experimental practices. Their outcomes are not predictable, and they can cause constant adjustments. With this concept of experiment, social analysts have a term that integrates the coping with surprises and the acknowledgment of ignorance. As people try out things, they better understand what they are doing as well as what they plan for the future, which is part of the recursive practice of experimenting. Since ignorance can never be delimited to zero, an experimental approach appears to be a useful long-term strategy for ecological design.

Immediately implementing the apparently "best" long-term management decision and then monitoring its effects usually results in only slight improvements in decisions over time. In contrast, what I call a *robust strategy*—a deliberate, well-designed experimental strategy that is open to surprises—can quickly merge into a superior long-term management decision. Drawbacks are to be expected, but an experimental design allows unexpected events to be absorbed without interrupting the whole process. If ecological design were to be understood as a stable and robust result, then it would be a closed system. Understood as an evolving iterative process, unexpected results and external influences (such as changing political conditions and natural alterations) can be perceived as a constructive parcel of strategic activity for successful and, in that sense, robust processes. The iterative process developed in chapters 2 and 3 and further illustrated in chapters 4 and 5 allows both positive and negative experiences (surprises) to be fed back into the next phase of the process. Understood in this way, the value of surprises has a pivotal position in the success and the process of innovation of the modern world since without surprises there would be no anticipatory joy, goal striving, or hope.

Whereas in long-term decisions with a belief in certainty in science, openness to surprises and acknowledgment of ignorance are very low, the chances that a robust process will emerge is also very low. Such a process can lead at best to a passive adaptation process and not an active experimental procedure that fosters surprises to learn from them. If we take the case of Montrose Point as a touchstone, the variation in the scale of openness to surprises and the acknowledgment of ignorance has been relatively small. However, the case of Leipzig's New Lake District, which moved from a strategy that welcomed surprises to a lock-in of development, can be seen as an example of how openness to surprises can be an important component of robust ecological design. More robustness can also lead to less preparedness due to a growing belief in certainty and safety. What then is needed is a radical lock-out of a stalemate situation, which allows moving forward. Sometimes such a lock-out can be brought about only by an external influence—that is, a total surprise (based on nescience) that makes obvious the actors' ignorance. This can be the prelude to a new phase of robustness.

There are limits to such strategies, and they can best be pursued on a local and perhaps regional scale with clearly defined boundaries of the experimental network so that surprising setbacks as well as benefits of the process and its options are unambiguous to many different parties. As has been shown here, this can help to engender trust among the actors involved and an acknowledgment of what is not known. When stakeholders are not able to oversee the learning from previous experiences and cannot agree on whether experimental procedures should be continued, an experimental approach does not appear to be a useful strategy. Furthermore, if a surprise cannot be clearly communicated as such by the actors involved (since not everyone has a taste for acknowledging unexpected turns as potentially useful), then an experimental strategy is doomed to failure.

In general, the call for "more action, less planning" needs to be complemented by concrete frameworks that have more to offer than mere ad hoc suggestions and the romanticizing of participation processes or open learning systems. In ecological restoration and landscape design projects, where the boundaries between research and implementation are permeable in both directions (mode 1 and mode 2 of science), an experimental framework that includes heterogeneous actors can be seen as a step to more robust restoration strategies. In such a model, social robustness always goes hand in hand with a high epistemic robustness since many projects in ecological design resemble a challenging form of cooperation between different forms of knowledge production. Robust strategies in ecological design need to be faced as an emergent practice based on iterative processes that allow learning *in and with* the ecosystem and human society. Given the complexity of interactions between human and nature systems and the highly unstable reactions that stem from natural and social systems, the integration of social preferences and scientific research results can best be handled by experimentally absorbing the unexpected.

To act experimentally, be less dependent on direct top-down local planning, and implement knowledge production, participants in planning on a regional basis need to regulate in a bottom-up fashion so that possible drawbacks can be addressed. The paradox here is that experimental approaches need strict regulation and cooperation at different regional levels so that preparedness for surprises is made possible. This means a shift away from mainly economic incentives and toward institutional capacities that help people to adapt to rapid and sometimes disorderly landscape changes and unforeseen events of the social system. The case of Montrose Point has shown that to restore a piece of land in a densely populated area, a governing agency, such as the United States Department of Agriculture's Forest Service, is needed to bring about surprises that help the stakeholders involved to learn and thus to make the project more socially *and* scientifically robust. The more experimental the character of activities on the ground, the more important state regulations become for coordinating innovative modes of experimental governance and public management. This statement runs counter to the debates on new forms of local governance and citizen organization, which proclaim the decline of the state and therefore of government (e.g., van Creveld 1999). Our cases, however, suggest that dealing with increasing ignorance as a normal part of contemporary decision making means that public experiments need to be well thought through, be tightly planned, and be helped by government regulations that empower committed actors in policymaking and implementation to learn reliably from surprises. They can be rendered as experimental niches that foster participation and learning to develop new ideas, new practices, and knowledge.

As was the case in Leipzig's New Lake District, ecological design projects with high expectations of planning tend to engage in major changes mainly after they have been confronted with negative surprises—that is, failure and crisis. These surprises may stem from external sources (such as extreme weather or the European Union's new political regulations), but they may also be generated deliberatively in the sense that they are waited for so that they may be learned from on a possibly small and thus manageable scale. Openness to surprises appears to have been evoked deliberatively in a recursive experimental design. The stakeholders involved can construct them either in response to an external constraint or in the absence of an external influence.

Both possibilities have frequently been the case in the design of Leipzig's New Lake District. In the first half of the 1990s, the Lake District resembled an almost perfect example of constructing and fostering surprises proactively for experimental practice, mainly because it had the most adventurous planners and they had nothing to lose. After brown coal mining ended and whole villages that had grown up because of mining were destroyed, a visionary "everything can be done" culture developed. Surprises were (knowingly or unknowingly) fostered for opportunistic practice. Whereas the lakeshore design in Chicago has taken place on landfill that piled up over more than a century, the design in Leipzig is taking place in an area that was completely changed (the soil was literally used up by mining over the course of some sixty years). In Chicago, a top-down strategy was radically turned into a bottom-up process that can be seen as a breakout of a lock-in situation. In Leipzig, the opposite happened between 1995 and 2005. This development portends to the limits of recursive practices that can lead to "lock-ins." Developments into the other direction, "lock outs," however, need to be expected. The lack of unified planning in and for the region and an overall vision for the future of the Lake District south of Leipzig had not taken on a feasible form. Indeed, it appeared that every lake's administration planned and "experimented" for itself. Only since 2003 has an overall framework for future planning of the New Lake District been in the development process, not as a framework to foster surprises and to develop robust processes but as a means for more top-down implementations that shuts any possibility of actively learning from surprising events.

Openness to surprises *alone* cannot expedite learning and lead to robust strategies. There is no universal panacea, and welcoming surprises does not make an exception. Worse, an experimental approach may easily turn into the perfect camouflage for failed projects: if surprises are good and yardsticks are to be developed over the course of experimentation, *any* project can claim to be a success. Robustness needs a recursive design and institutional embeddings that can register and further use the knowledge being added and handle the acknowledged nonknowledge. This needs to be accompanied by social support that believes that new knowledge can be produced and will be equipped with the potential to deliver the prerequisites for innovations, which provide advantage for social wealth and prosperity. It furthermore ought to be matched with the necessary prior knowledge to elevate experimental practice. Prior knowledge gained in different contexts can provide a platform for further interventions even if the respective platform is being superseded by new knowledge that renders obsolete earlier knowledge bases. In these stages (as in Montrose's early attempts via a top-down approach), accommodation of expertise took place largely through learning while doing.

Surprises are junctions for new directions in development and, in turn, can be supported or opposed but cannot be avoided. This "welcome imbalance" turns public experiments into a process and actors in this process into experimenters. This inescapable experimentation process can make the whole learning process recursive and thus robust—but also potentially endless.

Perspective: Surprises as Opportunity, Nonknowledge as a Working Base

If the interaction between modern science, society, and the natural world is framed as a real-world or public experiment, then practice must be based on recursiveness since knowledge production is part of the discovery process. Scientific activities in their context of application have certainly changed the processes via the implementation of "ecological experiments in the real world." The expectation of "surprises as opportunity" in ecological design projects and in that way that learning *in* ecosystems helps to relate scientific and societal activities to each other in a way that makes possible a wide and robust embedding and an embodiment with wider societal influences.

A final remark is in order on the role of the social scientist in these types of experimental practices. At first glance, in an experimental knowledge society that is focusing on experiments in society, the anthropologist, political scientist, geographer, or sociologist is sentenced to passive observation. However, one of the means of reflexivity in modern society has been the social sciences, perhaps most explicitly in the discipline of sociology. Sociology is generally rendered as a reflexive discipline that aims to develop an awareness of the social world (cf. Bourdieu 1990; Coleman 1990; Luhmann 1992). As a reflexive discipline, its subject matter also encompasses itself. In this vein, I would like to raise three points here. First, the historical discussion in chapter 2 demonstrates that from the very beginning of American social science, sociologists have imputed to society the language of experimenting. There can be no experimental practice without its reflexive description as experiment in terms of design, data collection, and interpretation of effects. Second, from a methodological point of view, the most consequential distinction between experiments in the laboratory and experiments in the real world is control versus lack of control with respect to boundary conditions and parameter variation. Researchers of the social world are not free to define at will the most feasible boundaries and parameter values. Instead, they have to take on decisions made by the responsible political bodies. Still, these are deliberate decisions that are activated by legislative or other institutional measures. Therefore, they can be taken as conditions of intervention to which different effects can causally be related. Even the best-controlled group experiments (for example, in correction institutions) carry features of an experimental society as manipulation of subjects or "victims" has (or should have) legal, ethical, and communicative limits. This leads to my third point—the involvement of participants. The standard method of experiment strictly detaches the experimenter from the experimental setting or system. Social scientists in a knowledge society of self-experimentation cannot possibly pursue this method. To a certain degree, all social groups from planners, ecologists, to concerned citizens are participant observers. Especially with the case of Montrose Point, there exists a broad variety of methodological models for negotiating and organizing participatory strategies.

In short, processes of science in public and the heterogeneous actors involved can lead to more effective reflexivity, and in this sense it means a relaxation toward more bottom-up approaches in knowledge production outside the laboratory. However, given the necessary preparedness for surprises, this relaxation needs to be planned and assessed even more carefully than traditional top-down approaches and strict substantive regulations. Planning the unplannable to learn from surprising events is much harder than trying to avoid surprises.

The framework presented above embraces an anticipatory approach to robust research implementations and learning. Although the surprises that are associated with ecological implementations cannot be eliminated, their negative effects may be made less painful and far shorter lived through experimental recursive practice incorporating the various modes of science, knowledge production, and implementations in public. The paradox result is that the more robust a strategy becomes, the more important surprising events (including negative ones—that is, failures and crisis) become as prerequisite for nonknowledge as a working base for the production of new and corroborated knowledge—that is, the foundation for social robustness.

As is discussed in the previous chapters, a good deal of modern science tends to extend research processes beyond the walls of the laboratory into wider society. The traditional model that used to describe the relationship between science and the public was one where the practical use of scientific knowledge was perceived in a linear and deductive fashion: research results are to be used by the public or policymakers to make decisions. These basic premises of scientific management have been questioned. The technical application of value-free, objective, and definite research results have long since been unmasked as at best an exception and more likely as a social fiction. Scientific research has moved out of the laboratory, taking with it in its train the uncertainty of knowledge production but also the opportunity to learn from surprises. This can be seen as a process of negotiation between science and the public.

In the way that ecological restoration appeared on the agenda as a mode 2 form of research in the late 1970s, later an official form of scientific knowledge production (mode 1) was gradually established before it began to be partially superseded by another phase of ascending mode 2 knowledge production. Thus, it can be concluded that in fields such as ecological design (and probably many more areas), there is a demanding interdependence between a mode 1 and the mode 2 production of knowledge. The analytical demarcation of forms of knowledge production does not split them off from one another but rather makes visible an increase of border traffic between them, which produces exactly the novel forms of cooperation and thus the important challenges that include new and surprising events that are located outside the sphere of quantifiable prediction and risk assessments. This means major challenges for democratic societies and their understanding of the role of science.

Recent debates in popular newspapers and magazines on the democratization of science and the call for a general change in science have reached the level of politics and public policies. Many of these debates circle around the dilemma of not knowing before an implementation whether the social and ecological outcomes will be acceptable to scientists as well as the public. This is an important point in ecological design. In many large-scale restoration and landscape renaturing projects in densely settled areas, political and scientific experts are often forced to play down the implicit experimental character of research-based restoration strategies by emphasizing rhetorics of safety because a possible loss of public acceptance is anticipated.

A final remark in this section refers to the goals for knowledge production outside of the realm of science and of the type of surprises-welcoming public experiments discussed above. It should be understood that, on the one hand, purely scientific interests would not suffice to foster one single surprising event affecting a single person or even a community. On the other hand, it is by no means impossible that knowledge interest, at least partially, coincides with public interests. Again, there is no general solution for reconciling differing interests and combining different modes of legitimization. However, there is always some scope for offering public participation and collective practice and gaining influence on defining experimental conditions and establishing scientific observation.

The core questions that have been explored in this study have focused on the societal and scientific prerequisites for successful ecological design projects in spite of ignorance on a landscape scale. All cases discussed have illustrated that the definition of a successful implementation can be developed during the implementation process itself, which means that goals have not been pursued via a master plan of action. In the examples discussed above—the attempts at controlling the spread of malaria, the restoration activities in an urban setting in Chicago, and the large-scale landscape transformations in eastern Germany—the experimental handling of ignorance and surprise has always helped to generate new and important sets of knowledge that hardly anybody had foreseen. Indeed, the results sometimes stood in contrast to the intentions and anticipations of some of the actors involved, but they also helped to produce a unique "edifice of knowledge" (Fleck 1979, 69).

Landscape design in a twenty-first-century knowledge society neither requires a narrow focus on purely economic incentives, safety, and scientific certainty nor can it be primarily built on issues such as the unemployment rate of a region or the potential of profits for investors. These would not be salient attributes to openness for surprising issues. Rather, openness to surprises in landscape development emphasizes, for example, the capacity of the actors and stakeholders involved to accept surprises as opportunities (as is discussed in the discovery process of certain acidification patterns in the water of Lake Cospuden). The cases discussed here have shown that experimental character comes about not always via rational decision making but sometimes through the gradual interpretation of novel events. New interpretations are based on the acknowledgment of ignorance through a challenge of routines, traditions, or accepted sets of knowledge. Daily routines can collapse because new knowledge has uncovered nescience and turned it into ignorance. Many observers argue that in the twenty-first century the potential for both progress and disruption is equally great since many processes of innovation can continue only by going beyond the limits of risk assessments and rationality. At least on local and regional levels, the use of experimental strategies that take seriously the inevitability of surprises and the uncovering of ignorance seems to be a way out of this dilemma. Hindsight does not reverse mistakes, but experimental strategies appear to be a useful avenue, especially in areas where the complexity of the situation and acknowledged ignorance do not allow decisions to be made based on clearly defined probabilities and certified expert knowledge. Thus, experimentally tinkering with ignorance and surprise is no romantic idea of learning by doing. It also differs from notions of muddling through, from incrementalism, or from mere adaptation measures. Experimenting in and with an ecological context needs to be designed as a socially acceptable and well-planned strategy to spark the unknown. This is no easy task, but experimenting with ignorance and surprise in coordinated practices that combine knowledge production and knowledge application appears to be a tolerable strategy on our path entering the knowledge society. Because the naturally occurring and human-induced ecological changes (such as global climate change) of this century will certainly leave their impressions on communities in very different forms, it is even more important to discuss what it takes to design a piece of land experimentally in the face of expected ignorance and surprise.

Notes

Chapter 2

1. All this amounts to what Ulrich Beck (1995) terms *organized nonliability*, where the burden for proving the culpability of an industrial actor tends to fall on the shoulders of underpowered individuals and where narrowing guilt down to one actor is impossible.

2. The popularity of criticizing the linear model of innovation and development is ongoing and ranges from stating that there never was a linear model (Edgerton 2004) to reconstructing different phases of the model (Godin 2006). In the following, I do not repeat the well-known criticisms of linear models but agree with Forman's (2007) observation that in official rhetoric a linear model might not exist anymore (if it ever did) but that at the level of belief it is still alive, although filed under different labels.

3. Another prominent example is Thorstein Veblen, exemplified especially in his programmatic article "Why Is Economics Not an Evolutionary Science?" (1898). Discussions on the unintended in classical economics can be traced back to at least the eighteenth century, however, and are perhaps most prominently illustrated in Adam Smith's reference to the invisible hand. Besides the earliest criticisms on the economic theories of the late nineteenth century from scholars such as Veblen (1898), twentieth-century critics attacked mainstream or neoclassical economics by pointing to the limits in economic theory to deal with issues of uncertainty and surprise in economic organization (e.g., Kay 1984; Shackle 1991).

4. Unless otherwise noted, this and the following quotes from German sources are my own translations.

5. This early debate was led mainly by Jokisch (1981), Wippler (1978), and van den Daele (1981). Besides the important work of Greshoff, Kneer, and Schimank (2003), authors such as Plé (1997) and Dietz (2004) have taken up the topic more recently.

6. In recent years, critics have questioned how and whether the concept of knowledge society in the original sense introduced by Bell and developed by Nico Stehr makes sense in regions outside of the so-called developed world. See, for instance, the discussions in Beerkens (2008), Gerke and Evers (2006), McElhinney (2005), Rohrbach (2007), and Saloma-Akpedonu (2008).

7. Following the work of Daniel Bell (1973), authors like Drucker (1973), Lane (1966), Lopata (1976), Machlup (1962), Price (1961), and the classical studies of Hayek (1945), Park (1940), and Znaniecki (1940) should be mentioned. For a discussion of some of these authors in relation to today's debates on the knowledge society, see Stehr (1994, chap. 1) as well as the contributions in Böhme and Stehr (1986). For an enlightening view on the development of the notion of postindustrialist society as the basis for a knowledge society, see Clark (2005).

8. The SER Policy Working Group developed this definition in 1996. The most recent version can be found online at http://www.ser.org/definitions.html.

9. The European Landscape Convention, which was initiated by the Congress of Regional and Local Authorities of the Council of Europe, is aimed at the conservation, management, and planning of all landscapes.

10. The more hard-science-oriented representatives of restoration ecology (Giardina et al. 2007) have challenged this view. For further discussions, see also Cabin (2007a), Ingram (2009), and Weinstein (2007). For a reconstruction of the history of ecological restoration and some of its challenging assertions toward the science of ecology, see Gross (2003a, 11–65).

11. There are more attempts to describe changing modes of knowledge making, such as the notion of *strategic science* (Martin and Irvine 1989; Rip 2004) or the classical thesis behind *action research* (for an overview and introduction, cf. Greenwood and Levin 1998), which also claimed (at least implicitly) that it pursues a better type of research, which in the future will gain in relevance. However, these observations have not been discussed nearly as much as the above-mentioned concepts regarding their relevance for a better understanding of contemporary science.

12. Based on a citation analysis, Hessels and van Lente (2008) show that the mode 2 thesis is the most widely accepted theory of the changing role of science in society.

13. A few contributions that have both positively as well as critically discussed the idea of a new mode of knowledge production include Beaulieu (2005), Godin (1998), Harloe and Perry (2004), Gross (2002), Harding (2008), Hessels and van Lente (2008), Huff (2000), Pestre (2000), Shinn (2005), and van Aken (2005), Weingart (1997). As an overview, see the contributions in Bender (2001).

14. When discussing nuclear hazards, Ulrich Beck put the contradiction between traditional experimental logic and real-world implementation decisively right: "Just as sociologists cannot force society into a test-tube, so technologists can only test nuclear reactors if they turn the world into a laboratory" (Beck 1995, 123).

15. About the only prominent field in sociology that uses an experimental approach that is modeled along the natural scientific laboratory experiment has been small-group research on behavioral change under stress and in a variety of settings under controlled conditions like in schools or correction institutions (cf. Bulmer 1986; Meeker and Leik 1995; Milgram 1974). For an analysis of different definitions of *experiment* in psychology textbooks compared to definitions in textbooks in physics, biology, and sociology from the 1930s to the 1970s, see Winston and Blais (1996).

16. The thought experiment has even been called the purest form of experimentation by some philosophers. The characterization of this special type of intellectual work as experimental, however, has been used only after the laboratory method of experiment was developed (cf. Gooding 1993; McAllister 2004).

17. Although Schulz (1970, 32) treats this type of experimentation outside the laboratory as the most recent form of experiment, he nevertheless names the utopian socialist Charles Fourier (1772–1837) as the originator of social experimentation. Unfortunately, neither Schulz nor Parthey and Wahl (1966, 231–240), who argue into a similar direction, elaborate this point any further. Zimmermann (1972), in his otherwise significant book, does not mention any of them.

18. For an overview of some of these studies in different fields, see Gross (2004a, 2009) and Gross and Krohn (2005).

19. On Georg Simmel's influence on Robert Park, see Gross (2001, 90–121).

20. These ideas are best illustrated by Dewey's lifelong commitment to extensive citizen participation in politics and his emphasis on public education as a means of achieving a more democratic society (Dewey 1988). On this issue in Dewey's work, see Brown (2009, chap. 5). As we have seen, these particular prescriptions draw not on pragmatism's symbolic interactionist legacy in the tradition of George Herbert Mead (1934) but from a distinctive theory of society. The symbolic interactionist theory is the source that Jürgen Habermas drew from in his conception of the individual actor who is involved in communicating through significant symbols (Habermas 1984). On the relationship between Habermas's theory and symbolic interactionism and the unexhausted possibilities of the two traditions, see Hinkle (1992).

21. Windelband was also Robert Park's dissertation adviser. On Park's time as a student in Germany and his studies with Windelband, see Gross (2001, chap. 5).

22. Perrow's observation was based on the inherent structural properties of technological systems, which, he believed, make accidents "inevitable, even 'normal'"
(Perrow 1984, 5).

23. Simmel's influence on American sociology has been well documented. However,
his ideas on the increase of unintended and surprising effects in modern society
have not been discussed. Although Simmel's influence was especially strong in the
early Chicago tradition of sociology and in its extensions in the works of Everett
Hughes, Louis Wirth, and Erving Goffman, several authors have analyzed Simmel's
exclusion from the canon of U.S. sociology after World War II, especially during the
Parsonsian hegemony (cf. Jaworski 1997; Levine 2000; Nichols 2001). For my own
evaluation of Simmel's ideas on unintended effects in comparison to his American
student Edward A. Ross, see Gross (2003b, 2004b).

24. There is much more to the technology push and demand cleavage than I am
able to indicate in this brief sketch. Its outline simply helps to throw into relief its
main contours to complete them with Simmel's ideas on science and technology.

25. In recent years, the social sciences have borrowed the notion of recursion from
computer programming and mathematics and focused on a repeated application of
a procedure to successive results of a process. In this sense, recursion is about a
process of well-defined self-reference (cf. Crozier 2007; Krohn 2007; Luhmann
2000).

26. As a positive example of such a surprise, we can refer to Kurt Wolff's reflections
on Simmel's idea of love for a certain landscape—the "surrender to a landscape,"
where the subjective love is repeated by the landscape itself (Wolff 2002, 44).

27. In this respect, a surprise is related to what Bauer (1969, 18) has labeled "second-
order consequences." The implicit differentiation between unintended and anticipated surprises can also be found in Bauer.

28. There have been debates about the true origin of the Thomas theorem ("If men
define situations as real, they are real in their consequences"), ever since the theorem
("the theorem of the definition of the situation," as it is also sometimes referred to)
first appeared in a book by William I. Thomas and his wife, Dorothy Swaine Thomas
(Thomas and Thomas 1928, 571–572). However, as Robert Merton verified with the
help of Dorothy Swaine Thomas, the concept of defining the situation was strictly
W.I.'s. (Merton 1995).

29. Although I use the words *unanticipated* and *unexpected* interchangeably throughout the book, in many cases both words cannot be meaningfully established as
synonyms. For instance, you normally say that you have set aside extra beers for
unexpected guests, not for unanticipated guests.

30. Jasanoff (2005) discusses several traits of civic epistemologies. In this book, I am
especially interested in the forms of participation that are relevant to knowledge

production in public, public accountability, and trust in expertise in face of igno-rance. Jasanoff (2005, 255–259) lists six dimensions—styles of public knowledge making, public accountability, demonstration practices of knowledge claims, regis-ters of objectivity, accepted bases of expertise, and visibility of expert bodies.

31. This parallels the reintroduction process in North America, where the Defenders of Wildlife (http://www.defenders.org) dealt with the economic concerns of the local ranching industry for the loss of livestock. Financed by private donors, Defenders of Wildlife established a trust that compensates ranchers for verified livestock losses to wolves so that the economic burden is shifted from ranchers to the proponents of wolves.

32. Paul Gobster (2007) has made a convincing case that many restoration projects resemble a "museumification" of nature, which excludes the wildness and thus naturalness of nature.

Chapter 3

1. Ryle used the example of a boy playing chess. He suggested that the boy can be said to know how to play chess if his action displays the rules of chess even if he cannot recite the rules. This is similar to Polanyi's notion of tacit knowledge (1958).

2. Even if knowledge existed as an objective given, for a social scientist it can be empirically registered only by interpreting these different activities of individuals and groups.

3. The literature on complexity theory and complex systems in the context of chaos theory and the theory of self-organizing systems is immense, so I cannot review it here. For interesting applications, see the classical studies by Karl Weick (2001) on loose couplings and more recent discussions of health care management (e.g., McDaniel, Jordan, and Fleeman. 2003) and innovation networks (e.g., Küppers 2002).

4. Smithson, who for some fifteen years has worked in areas other than theorizing ignorance, has started working in this area again. Referring to his 1989 book, he concludes that real progress on interdisciplinary work on ignorance has not been made (Smithson 2008, 226). However, see the volume by Proctor and Schiebinger (2008), which includes a new Smithson essay. This book indicates a growing interest in the topic among historians.

5. To add even more terminological confusion, in a revised English version of Beck's essay on nonknowledge, *Nicht-Wissen* is translated as *unawareness* (cf. Beck 1999, 109–132).

6. In a translation by Kurt Wolff (Simmel 1964, 307–376) first published in 1950, *Nichtwissen* is uniformly translated as *ignorance*, with one exception, where Wolff

uses the term *nonknowledge* (ibid., 312). It is not clear why Wolff changed this here. In some of his own writings, Wolff uses an interesting notion of *nonknowledge* meaning a residual area of other forms of knowledge covering "all that might be there but is not" (Wolff 1943, 121). This meaning is different from Simmel's *Nichtwissen*.

7. The belief that nothing is knowable led to agnosticism, the doctrine that certainty about any absolute truth is unattainable. Thomas Huxley introduced the term *agnostic* in 1869 to describe his philosophy, which rejects all claims to spiritual or mystical knowledge (cf. Lightman 1987).

8. Another reason for refraining from using the terms *known* and *unknown unknows* is that they have become rather infamous since Donald Rumsfeld, U.S. secretary of defense from 2001 to 2006, used them at a defense department briefing in 2002. The terms were used in pop culture to mock Rumsfeld and his political environment.

9. Some authors, like Fritz Machlup (1962, 16), have used the term *nonknowledge* in the English language to indicate something that is not knowledge (e.g., assumptions or beliefs). In this sense, see also its usage in the literature on service economies (e.g., Ducatel 2000).

10. The quotes by Simmel in this section are my translations from the German based on the original translation by Albion Small (Simmel 1906).

11. Giddens (1990) adds the notion of confidence. He uses the term *confidence* to describe the faceless commitments that rely on the professional fulfillment of expertise. Since this type of facelessness is rarely found in ecological design, where personal expertise is the most important factor for dealing with ignorance, I do not consider this term any further. For a similar result but in reference to the concept of professionalism and its relation to trust, see Svensson (2006).

12. In a similar vein, Heinrich Popitz (1968) has stressed the importance of *Nichtwissen* as a preventive action and a maneuver in treating delinquent behavior.

13. Although the concept of bounded rationality appears to be helpful for this discussion, its challenges for economic theory are most often stated as an unfortunate issue of economic life and not (as we view ignorance and surprise here) as something to be potentially welcomed.

14. More generally, Stocking (1998, 177) calls for the establishment of a sociology of scientific ignorance (SSI) to complement and expand the traditional sociology of scientific knowledge (SSK). Earlier, Stocking and Holstein (1993, 187) defined *ignorance* as "absence of knowledge, and uncertainty, incompleteness, bias, error, and irrelevance."

15. Knight (1921) developed a well-known distinction between *risk* and *uncertainty*: risk can be calculated, but uncertainty cannot. Some researchers still use *risk* and

uncertainty interchangeably. Here I try to avoid the term *uncertainty* since it has a clearly negative connotation. For an excellent overview on recent interdisciplinary work on risk and related fields, see Taylor-Gooby and Zinn (2006).

16. Over the last decade, Luhmann's sociology of risk has received more attention and criticism than I can discuss here, but see Rosa (2003) or Wehling (2006) and the literature listed therein. Most generally, Renn (2008, 32) has argued that although Luhmann's systems theory provides explanations for why certain contested decisions on risks fail, it "offers few clues of how contemporary societies reach closure on collectively binding standards or decisions," although "it is obvious that such decisions are made routinely."

17. At the risk of overstressing Simmel's insights, I want to mention what he wrote about this issue: "Whatever quantities of knowledge and nonknowledge must commingle, to make possible the detailed practical decision based on trust, will be determined by the historic epoch, the ranges of interests, and the individuals" (Simmel 1992, 393–394; cf. Simmel 1906, 450).

18. In her 1999 book, Knorr Cetina seemed to use the term *negative knowledge* interchangeably with *liminal knowledge*. In a more recent essay, she says that negative knowledge "is gained from the disturbances, distortions, errors and uncertainties of research" (Knorr Cetina 2007, 366). Liminal knowledge, in turn, is derived out of what cannot be studied via the identification of domains "by measuring the properties of the objects that interfere with them and distort them" (ibid.). Machlup (1980, 144ff.) uses a different notion of *negative knowledge* that I do not pursue further because it covers too many overlapping meanings. Among his thirteen possible usages of the term *negative knowledge*, for example, Machlup grouped sets of knowledge that have turned out to be erroneous, vague, rejected, alternative, obsolete, or controversial.

19. Faber and Proops (1998) offer a helpful taxonomy of types of ignorance, which are compatible with issues discussed here. However, their work is focused on long-run interactions between the economy and the environment and their treelike taxonomy appears to be unidirectional with no temporal, conceptual, or causal reconnection between different types of unknowns. Similarly, see also Kerwin's (1993, 178) or Smithson's taxonomy of ignorance (1989, 9).

20. The term *nescience* is difficult to translate into German. The possible translations as *Unwissen* and *Unwissenheit* point to what cannot be known but also have a clearly negative connotation (e.g., a person is too dumb too know any better).

21. At first sight, the term *extended knowledge* means nothing more than "knowledge at time $t + 1$," and thus a new term is not needed. However, *extended knowledge* can as well mean "knowledge at time $t + n$," which also includes a connectivity to new ignorance, new nonknowledge, and so on.

22. Over the last decade, I have tried several variants of this scheme in different cases and with different terminologies. For some of these cases, see Gross, Hoffmann-Riem, and Krohn (2003) and Gross (2007b). For further usages or criticism see Frickel et al. (2010), Wehling (2006), Wibeck (2009), and Williams (2008).

23. Many deviant results, at least after a certain amount of time, are the result of procedural errors and not real surprises in the behavior of the system. Kai Lee (1993, 66) suggests that experimentation "helps to create a social system able to recognize the needle of real surprise in the haystack of mundane error."

24. Walters (1986, 9) saw the value of adaptive management in questioning some basic management assumptions and saw adaptive management as able to (1) group management problems and recognize practical constraints, (2) represent knowledge in models of dynamic behavior that identify assumptions and predictions so that experience can further learning, (3) represent uncertainty and identify alternate hypotheses, and (4) design policies to provide continued resource productivity and opportunities for learning.

25. See, for instance, the classical criticism by Etzioni (1967) and Forester (1984).

Chapter 4

1. There are many more differences between ecological restoration and conservation biology. For instance, the latter has organisms as its major focus and is less focused on entire ecosystems and their functions. However, those differences will not concern us here.

2. Consider Emile Durkheim (1933, 348), who wrote: "As for the physical world, since the beginning of history it has remained sensibly the same, at least if one does not take account of novel ties which are of social origin." Following Durkheim, social scientists can bracket out any outside powers of nature as an explanatory variable for social explanations. For further discussions on this issue in Durkheim's writings, see Gross (2001), Baerlocher and Burger (2010), and Rosa and Richter (2008).

3. I owe the example of public health research to Kelly Moore.

4. Data on Montrose Point come from different sources, including minutes and transcripts of focus-group meetings made available by Paul Gobster and Lynne Westphal, the observation of activities at the site, published materials, evaluations of archive materials, and fifteen semistructured interviews with stakeholders. My first field trip to the site was with Paul Gobster and Reid Helford in the spring of 2000. Although I visited Montrose Point a few more times in the following years, in the fall of 2003 I spent a full semester at Loyola University to do research for this book, and the school is almost in walking distance to the site. I began my work by

interviewing decision makers and stakeholders who were previously involved in different restoration projects in the Chicago area to get a general understanding of the field. I conducted an analysis of processes based on uttered acknowledgment of what was not known by using my conceptual framework on ignorance and surprise to shed new light on how those processes inform existing scholarly knowledge on dealing with ignorance. I initially codified much of the written material with a computer program for qualitative text analysis during a research project at Bielefeld University. I later used the standard methodology of content analysis to identify text passages to allocate them to predefined themes and categories based on the theoretical concept of nonknowledge and ignorance. My last stay in Chicago was in February and March 2008 to finalize my research on the development of the site and to verify my analysis by comparing findings from additional research sites in the Chicago area.

5. This coalition is a volunteer-driven effort consisting of four teams as a steering committee. The group's steering committee includes representatives from the Chicago Park District, the City of Chicago Department of the Environment, the Lake County Forest Preserves, the Indiana Dunes Environmental Learning Center, the Indiana Dunes National Lakeshore, the National Audubon Society, and the John G. Shedd Aquarium. The four teams are the scientific team, a land management team, an office for education and outreach, and a policy and planning team. A steering committee of executives oversees the initiatives, including approval of budgets and projects. The Chicago Region Biodiversity Council encompasses the chief executives of all member institutions. The executive members elect the steering committee. A proposals committee solicits and reviews proposals for priority conservation projects and recommends funding; the committee has one or two representatives from each of the four teams (cf. Alario 2000; Moskovits, Fialkowski, Mueller, and Sullivan 2002).

6. In a program sponsored by Friends of the Parks in partnership with the Chicago Public Schools (CPS), the Illinois Department of Natural Resources (DNR), the Chicago Audubon Society, and the Chicago Corinthian Yacht Club, Nature along the Lake (NATL) uses the nature preserve at Montrose Point as a nature classroom.

7. The Ecological Rehabilitation Plan from the Chicago Park District lists seven major plant species and a few dozen others.

8. *Baby dune* is not a technical term. At several field trips and site visits to the point, most notably those I undertook with Paul Gobster, Alanah Fitch, and Lynne Westphal, the emerging piles of sand have been referred to as baby dunes. The *Chicago Tribune* also used the term occasionally.

9. In the summer of 2004, the Illinois Department of Natural Resources (IDNR) tried to control invasive species at the site and to preserve the site's unique flora and fauna, including uncommon geological features such as a *panne* (wet depression in

a swale system). IDNR subsequently nominated the site for INAI status, which opens the way for much greater enhancement of Montrose Beach Dunes.

10. Barbara Adam (1998) is perhaps the best-known social scientist to point to the importance of ecological time. She sees human engagement with nature in terms of interlocking cycles of social and ecological time.

11. Kim's article on organizational learning in the automobile industry (Kim 1998) was the inspiration for my figure 4.2 and its approach to relating ignorance and surprise to the notion of robustness. Kim deals with a completely different topic (car manufacturing), however, and instead of integrating ignorance and surprise, he refers to the factors of "prior knowledge base" and "intensity of effort" as prerequisites for what he calls *absorptive capacity*.

Chapter 5

1. The term *surface mining* refers to any mining activity that is performed at or near the earth's surface. *Strip mining* normally refers to the surface mining of coal rather than metalliferous ores. *Open-pit* and *open-cast mining* are terms that are sometimes used to refer to the removal of ores and minerals near the surface to distinguish the surface mining of metalliferous ores from other types of surface mining, such as the strip mining of coal. However, I use the terms *strip mining* and *open-pit mining* interchangeably in this chapter since in the literature both terms are used for surface mining activities.

2. Today the terms *brown coal* and *lignite* are used synonymously. The etymology of the word *lignite* can be traced to the Latin *lignum* ("wood"). It is coal that shows traces and some of the texture of the wood that it once was. Until the 1950s, a distinction was made between a more consolidated material (lignite) and a less consolidated, friable material (brown coal). Furthermore, brown coal and lignite have also been distinguished in their moisture content: brown coal has more moisture and lignite less. Depending on the source, what is called brown coal in one instance could be called lignite in the another. Lignite and brown coal generally have a higher moisture content (up to 45 percent) and a high sulfur content compared to other types of coal. For reasons of uniformity, I use the term *brown coal* throughout this chapter except in quotes where the term *lignite* is used.

3. Research for this chapter began in April 2005, although I had collected information on the New Lake District before this date. I collected historical texts and news clippings and conducted twenty-five semistructured interviews with local residents, planners, lake scientists, and administrators. I taped interviews and took detailed field notes. All interviewees were aware of my status as a researcher from an academic setting, and many knew colleagues at my current employer, which probably made my entry into the field easier. The interviews that I quote from in this chapter

took place mainly between July 2005 and March 2007. As in the previous chapter on Montrose Point, I complemented the interviews with other materials that I collected from planning meetings, work sessions, field trips, and various planning offices (including minutes and in-house publications). The material then was codified with a qualitative text analysis program to organize data and identify recurring themes. Not all interviewees felt comfortable having me attribute their comments on them personally; so most people that I interviewed remain anonymous. Unlike otherwise indicated, all statistics and other historical data are extracted from internal materials from the Regional Planning Department of West Saxony.

4. As Rink (2002, 88) observed, it is generally difficult to judge socialist Germany's "socialist" environmental policy: "The excellent legislation and institutionalization remained largely ineffective, and the positive impacts . . . were in fact unplanned side-effects of economizing and self-sufficiency measures."

5. As many of the actors involved in the early 1990s clarify in interviews that I conducted between 2005 and 2007, the knowledge basis for the recultivation and flooding of former strip-mining areas, in the German Democratic Republic has been quite large (cf. Krummsdorf and Krümmer 1981). At the same time, however, many political and economic necessities lead to fundamental gaps in knowledge, which also led to the environmental situation in the early and mid-1990s.

6. However, unlike opposition movements against open-pit mining operations in North America (e.g., Montrie 2003; Shnayerson 2008), the movement in eastern Germany did not begin until the late 1980s and never reached a similar influence. For a general overview and comparison between eastern and western German environmental organizations and their effect on policy making, see Markham (2008).

7. Including the Lusatian region, forty-five lakes with a total volume of 4.2 billion cubic meters of water will be created in an area of 2,120 square kilometers. Some of these lakes will be among the largest in Germany.

8. I first encountered this description in June 2005 in a conversation with Karl-Detlef Mai, but it has been used in official documents such as those of the State Office for the Environment and Geology of Saxony.

9. The Lausitzer und Mitteldeutsche Bergbau- und Verwaltungsgesellschaft (Lusatian and Central German Mining Administration Company, LMBV) is a German, federally owned management company that is responsible for project organization in areas such as the revitalization of postmining landscapes. By the end of the first decade of the twenty-first century, the LMBV had been involved in the design of embankments, the filling of low areas, and the demolition of unused factory buildings. In coming decades, the main focus of its work is going to be the rehabilitation and monitoring of water balances, flooding control, and the recultivation of new surfaces (Schlenstedt and Bender 2004).

10. This is in accordance with the Water 'Framework Directive of the European Union, which calls for the use of the groundwater that is pumped out of open-pit mines prior to coal extraction (cf. Linke and Schiffer 2002).

11. In 2004, 8.5 million tons of brown coal were excavated in the two still active open-pit mines south of Leipzig. This is a miniscule quantity compared to the amounts that were mined prior to 1990, but it was still more than was mined in all other regions in Germany.

12. Despite extensive historical investigations, it is not even clear if the heavy metals in the overburden heaps might have had *geogenic* origins (that is, based on factors that originate naturally in the soil) rather than *anthropogenic* origins (that is, caused by humans).

13. This observation is supported by an analysis of clippings from the region's major daily and weekly newspapers on the development of mining pits, as well as interviews with stakeholders of the planning and implementation process.

14. A *reclaimer* (*Absetzer*) is a large, earth-moving technological machine that dumps the overburden from mining activities to rebuild the surface of the landscape.

15. Miller (2000, 281) summarized: "I mistrust the philosophic tradition that seeks to link courage too closely with reason, thereby reducing courage to prudence's handmaid."

16. This observation is implicitly supported by Shrum, Genuth, and Chompalow (2007). In their study on scientific collaborations, they could not support the thesis that research projects that formed from preexisting relationships had any overall advantage in terms of trust. Perhaps, as they summarize, "trust is a more complex issue" (Shrum et al. 2007, 157).

17. In less than one minute, more than 1 million cubic meters of soil can move, even in areas that received the rubble many decades earlier.

18. The term *flash flooding* is normally used to describe a rapid flooding of low-lying areas, rivers, and streams that is caused by the rapid rainfall associated with a thunderstorm. The actors in the development of Leipzig's New Lake District used the term to distinguish their flooding speed from regular open-pit flooding procedures.

19. The flood that was produced by the Elbe and adjacent rivers in the Czech Republic and eastern Germany in August 2002 was referred to as the "flood of the century" in Germany.

20. Especially in 2007, when state sponsorship for the construction of forty new summer residences as part of a new holiday village was approved, public debates on the marketability of such houses were held in outlets such as the Leipziger Volkszeitung (LVZ).

21. Although population decline is extreme in regions such as eastern Germany, Italy, and many eastern European countries, on a global scale a reduction cannot be expected before the second half of the twenty-first century (cf. Wattenberg 2004).

22. In face of global warming, one could speculate that this New Lake District will become as attractive to national and international tourists as is the old Lake District in Great Britain or even the tourist centers in other parts of Europe. Current trends to vacation in central and western Europe instead of southern European countries might support such a development. However, this would certainly cause other major environmental problems, such as the negative environmental effects of mass tourism.

23. This was the meeting (Konzept Leitbild im Zuge der Gesamtfortschreibung des Regionalplans Westsachsen) that was held on March 16, 2006. On the general challenge of coordinating the new lake districts all over eastern Germany, see Krüger, Kadler, and Fischer (2002).

24. The metaphor that the energy input in a motor boat equals an investment in landscape development can certainly be questioned. However, the point that Oswalt et al. (2002) have made should be clear.

25. The chief purpose of participation in Germany is to support an information flow that makes people aware of public issues. As Jasanoff (2005) has noted, it often is found in insulated regulatory processes in Germany. Although many of the stakeholders that I interviewed felt that they are not really involved in any decision-making process, they rarely saw this as a problem.

References

Abel, Andrea, Anne Michael, Angelika Zartl, and Florian Werner. 2000. Impact of Erosion-Transported Overburden Dump Materials on Water Quality in Lake Cospuden Evolved from a Former Open-Cast Lignite Mine South of Leipzig, Germany. *Environmental Geology* 39 (6): 683–688.

Abels, Gabriele, and Alfons Bora. 2004. *Demokratische Technikbewertung*. Bielefeld, Germany: Transcript.

Adam, Barbara. 1998. *Timescapes of Modernity: The Environment and Invisible Hazards*. London: Routledge.

Agrawal, Arun, and Clark C. Gibson, eds. 2001. *Communities and the Environment: Ethnicity, Gender, and the State in Community-based Conservation*. New Brunswick, NJ: Rutgers University Press.

Ahmed-Ullah, Noreen S. 2007. Nature Thrives in Heart of City. *Chicago Tribune*, May 27, 5 (Metro section).

Ahmed-Ullah, Noreen S. 2008. Fences Upset Park Volunteers: Many Helpers Peeved They Weren't Asked First. *Chicago Tribune*, July 2, 4 (Metro section).

Alario, Margarita. 2000. Urban and Ecological Planning in Chicago: Science, Policy and Dissent. *Journal of Environmental Planning and Management* 43 (4): 489–504.

Altmoos, Michael. 1999. *Systeme von Vorranggebieten für Tierarten-, Biotop- und Prozessschutz: Auswahlmethoden unter Einbeziehung von Habitatmodellen für Zielarten am Beispiel der Bergbaufolgelandschaft im Südraum Leipzig, Bericht No. 18/99*. Leipzig: Helmholtz Centre for Environmental Research – UFZ, Germany.

Anderies, John M., Marco A. Janssen, and Elinor Ostrom. 2004. A Framework to Analyze the Robustness of Social-Ecological Systems from an Institutional Perspective. *Ecology and Society* 9 (1): 18. URL: http://www.ecologyandsociety.org/vol9/iss1/art18/ (last accessed May 2009).

Arendt, Hannah. 1959. *The Human Condition*. Chicago: University of Chicago Press.

Argyris, Chris, and Donald A. Schön. 1978. *Organizational Learning: A Theory of Action Perspective*. New York: McGraw-Hill.

Arnstein, Sherry R. 1969. A Ladder of Citizen Participation. *Journal of the American Institute of Planners* 35 (4): 216–224.

Aronson, James, Sue J. Milton, James N. Blignaut, and Andre F. Clewell. 2006. Nature Conservation as If People Mattered. *Journal for Nature Conservation* 14 (3–4): 260–263.

Baehr, Peter. 2002. Identifying the Unprecedented: Hannah Arendt, Totalitarianism, and the Critique of Sociology. *American Sociological Review* 67 (6): 804–831.

Baerlocher, Bianca, and Paul Burger. 2010. Ecological Regimes: Towards a Conceptual Integration of Biophysical Environment into Social Theory. In *Environmental Sociology: European Perspectives and Interdisciplinary Challenges*, ed. Matthias Gross and Harald Heinrichs. Dordrecht: Springer.

Baldwin, A. Dwight, Judith de Luce, and Carl Pletsch, eds. 1994. *Beyond Preservation: Restoring and Inventing Landscapes*. Minneapolis: University of Minnesota Press.

Bauer, Henry A. 2001. Anomalies and Surprises. *Journal of Scientific Exploration* 15 (4): 459–463.

Bauer, Raymond A. 1969. *Second-Order Consequences: A Methodological Essay on the Impact of Technology*. Cambridge, MA: MIT Press.

Beaulieu, Lionel J. 2005. Breaking Walls, Building Bridges: Expanding the Presence and Relevance of Rural Sociology. *Rural Sociology* 70 (1): 1–27.

Beck, Ulrich. [1986] 1992. *The Risk Society: Towards a New Modernity*. London: Sage.

Beck, Ulrich. [1988] 1995. *Ecological Politics in an Age of Risk*. Cambridge: Polity Press.

Beck, Ulrich. 1996. Wissen oder Nicht-Wissen? Zwei Perspektiven reflexiver Modernisierung. In *Reflexive Modernisierung*, ed. Ulrich Beck, Anthony Giddens, and Scott Lash, 289–315. Frankfurt am Main: Suhrkamp.

Beck, Ulrich. 1999. *World Risk Society*. Oxford: Polity Press.

Beck, Ulrich. 2000. The Cosmopolitan Perspective: Sociology of the Second Age of Modernity. *British Journal of Sociology* 51 (1): 79–105.

Beck, Ulrich, Wolfgang Bonss, and Christoph Lau. 2003. The Theory of Reflexive Modernization: Problematic, Hypotheses and Research Programme. *Theory, Culture and Society* 20 (2): 1–33.

Beck, Ulrich, Anthony Giddens, and Scott Lash, eds. 1994. *Reflexive Modernization: Politics, Tradition and Aesthetics in the Modern Social Order.* Stanford: Stanford University Press.

Becker, Howard S. 1994. "Foi por Acaso": Conceptualizing Coincidence. *Sociological Quarterly* 35 (2): 183–194.

Beerkens, Eric. 2008. University Policies for the Knowledge Society: Global Standardization, Local Reinvention. *Perspectives on Global Development and Technology* 7 (1): 15–36.

Bell, Daniel. 1973. *The Coming of Post-industrial Society: A Venture in Social Forecasting.* New York: Basic Books.

Bender, Gerd, ed. 2001. *Neue Formen der Wissenserzeugung.* Frankfurt am Main: Campus Verlag.

Berkes, Fikret, and Nancy J. Turner. 2006. Knowledge, Learning and the Evolution of Conservation Practice for Social Ecological System Resilience. *Human Ecology* 34 (4): 479–494.

Berkner, Andreas. 2000. The Lignite Industry and the Reclamation of Land: Developments in the Rhenish, Central German and Lusatian Mining Areas since 1989. *Beiträge zur Regionalen Geographie* 52: 186–201.

Berkner, Andreas. 2001. Geburt eines Sees. *Südraum Journal* 12: 50–55.

Berkner, Andreas. 2004. Wiedernutzbarmachung im Braunkohlebergbau für forschungsseitige Grundlagen. *Zeitschrift für Angewandte Umweltforschung. Sonderheft* 14: 217–227.

Bernstein, Jay H. 2009. Nonknowledge: The Bibliographical Organization of Ignorance, Stupidity, Error, and Unreason: Part One. *Knowledge Organization* 36 (1): 17–29.

Bittlingmayer, Uwe H. 2005. *Wissensgesellschaft als Wille und Vorstellung.* Konstanz: Univeritätsverlag Konstanz.

Böhme, Gernot, and Nico Stehr, eds. 1986. *The Knowledge Society: The Growing Impact of Scientific Knowledge on Social Relations.* Dordrecht: Reidel.

Bontje, Marco. 2004. Facing the Challenge of Shrinking Cities in East Germany: The Case of Leipzig. *GeoJournal* 61 (1): 13–21.

Borowski, Ilke, and Claudia Pahl-Wostl. 2008. Where Can Social Learning Be Improved in International River Basin Management in Europe? *European Environment* 18 (4): 216–227.

Böschen, Stefan. 2000. *Risikogenese. Prozesse gesellschaftlicher Gefahrenwahrnehmung: FCKW, Dioxin, DDT und Ökologische Chemie.* Opladen: Leske + Budrich.

Böschen, Stefan, and Peter Wehling. 2004. *Wissenschaft zwischen Folgenverantwortung und Nichtwissen*. Wiesbaden: Verlag für Sozialwissenschaften.

Botkin, Daniel B. 1990. *Discordant Harmonies: A New Ecology for the Twenty-first Century*. Oxford: Oxford University Press.

Bourdieu, Pierre. 1990. *In Other Words: Essays towards a Reflexive Sociology*. Palo Alto: Stanford University Press.

Bradshaw, Andrew D. 1993. Ecological Restoration as Science. *Restoration Ecology* 1 (2): 71–73.

Bradshaw, Andrew D. 1994. The Need for Good Science: Beware of Straw Men. Some Answers to Comments by Eric Higgs. *Restoration Ecology* 2 (3): 147–148.

Brand, Fridolin S., and Kurt Jax. 2007. Focusing the Meaning(s) of Resilience: Resilience as a Descriptive Concept and a Boundary Object. *Ecology and Society* 12 (1): 23. http://www.ecologyandsociety.org/vol12/iss1/art23/ (last accessed May 2009).

Brand, Karl-Werner. 2010. Social Practices and Sustainable Consumption: Benefits and Limitations of a New Theoretical Approach. In *Environmental Sociology: European Perspectives and Interdisciplinary Challenges*, ed. Matthias Gross and Harald Heinrichs. Dordrecht: Springer.

Brecht, Bertolt. 1980. *The Life of Galileo*. London: Methuen.

Briggle, Adam. 2008. Review: Questioning Expertise. *Social Studies of Science* 38 (3): 461–470.

Brint, Steven. 2001a. *Gemeinschaft* Revisited: A Critique and Reconstruction of the Community Concept. *Sociological Theory* 19 (1): 1–23.

Brint, Steven. 2001b. Professionals and the "Knowledge Economy": Rethinking the Theory of Postindustrial Society. *Current Sociology* 49 (4): 101–132.

Brooks, Harvey. 1986. The Typology of Surprises in Technology, Institutions, and Development. In *Sustainable Development of the Biosphere*, ed. William C. Clark and R. E. Munn, 325–348. Cambridge: Cambridge University Press.

Brown, Mark B. 2009. *Science in Democracy: Expertise, Institutions, and Representation*. Cambridge, MA: MIT Press.

Bulmer, Martin. 1986. Evaluation Research and Social Experimentation. In *Social Science and Social Policy*, ed. Martin Bulmer, 155–179. London: Allen and Unwin.

Cabin, Robert J. 2007a. Science and Restoration under a Big, Demon Haunted Tent: Reply to Giardina et al. (2007). *Restoration Ecology* 15 (3): 377–381.

Cabin, Robert J. 2007b. Science-Driven Restoration: A Square Grid on a Round Earth? *Restoration Ecology* 15 (1): 1–7.

Calhoun, Craig, ed. 2007. *Sociology in America: A History.* Chicago: University of Chicago Press.

Callaway, Ewen. 2007. Malaria Research Should Go "Back to Basics." *Nature* 449:266.

Campbell, Donald T. 1988. *Methodology and Epistemology for Social Science: Selected Papers.* Chicago: University of Chicago Press.

Carlson, Jean M., and John Doyle. 2002. Complexity and Robustness. [PNAS] *Proceedings of the National Academy of Sciences of the United States of America* 99 (suppl. 1): 2538–2545.

Carrier, Martin, Don Howard, and Jeff Kourany, eds. 2008. *The Challenge of the Social and the Pressure of Practice: Science and Values Revisited.* Pittsburgh: University of Pittsburgh Press.

Chew, Sing C. 2008. *Ecological Futures: What History Can Teach Us.* Lanham, MD: Alta Mira Press.

Chicago Park District, ed. 1991. *Lincoln Park Restoration and Management Plan: Historic Preservation Analysis.* Chicago: Office of Research and Planning.

Clark, Terry N. 2005. Who Constructed Post-Industrial Society? An Informal Account of a Paradigm Shift at Columbia, Pre-Daniel Bell. *American Sociologist* 36 (1): 23–46.

Coe-Juell, Lindy. 2005. The Fifteen-Mile Reach: Let the Fish Tell Us. In *Adaptive Governance: Integrating Science, Policy, and Decision Making,* ed. Ronald W. Brunner, Toddi A. Steelman, Lindy Coe-Juell, Christina M. Cromley, Christine M. Edwards, and Donna W. Tucker, 47–90. New York: Columbia University Press.

Coleman, James S. 1990. *Foundations of Social Theory.* Cambridge, MA: Harvard University Press.

Collingridge, David. 1983. Hedging and Flexing: Two Ways of Choosing Under Ignorance. *Technological Forecasting and Social Change* 23 (2): 161–172.

Collier, Stephen J., and Andrew Lakoff. 2008. Distributed Preparedness: The Spatial Logic of Domestic Security in the United States. *Environment and Planning. D, Society & Space* 26 (1): 7–28.

Collins, Harry M. 1988. Public Experiments and Displays of Virtuosity: The Core-Set Revisited. *Social Studies of Science* 18 (4): 725–748.

Collins, Harry M., and Robert Evans. 2007. *Rethinking Expertise.* Chicago: University of Chicago Press.

Constant, Edward W. 2000. Recursive Practice and the Evolution of Technological Knowledge. In *Technological Innovation as an Evolutionary Process*, ed. John Ziman, 219–233. Cambridge: Cambridge University Press.

Cook, Thomas D., and Donald T. Campbell. 1979. *Quasi-Experimentation: Design and Analysis Issues for Field Settings*. Boston: Houghton Mifflin Company.

Cooke, Bill, and Uma Kothari, eds. 2001. *Participation: The New Tyranny?* London: Zed Books.

Couch, Chris, Jay Karecha, Henning Nuissl, and Dieter Rink. 2005. Decline and Sprawl: An Evolving Type of Urban Development—Observed in Liverpool and Leipzig. *European Planning Studies* 13 (1): 117–136.

Covington, William W. 2003. Restoring Ecosystem Health in Frequent-fire Forests of the American West. *Ecological Restoration* 21 (1): 7–11.

Crozier, Michael. 2007. Recursive Governance: Contemporary Political Communication and Public Policy. *Political Communication* 24 (1): 1–18.

Davis, Mark A., and Lawrence B. Slobodkin. 2004. The Science and Values of Restoration Ecology. *Restoration Ecology* 12 (1): 1–3.

De Soto, Hermine G. 2000. Contested Landscapes: Reconstructing Environment and Memory in Postsocialist Saxony-Anhalt. In *Altering States: Ethnographies of Transition in Eastern Europe and the Former Soviet Union*, ed. Daphne Berdahl, Matti Bunzi, and Martha Lampland, 96–113. Ann Arbor, MI: University of Michigan Press.

Dettmar, Jörg. 2004. Neue Landschaften—Verpasste Chancen. *Garten + Landschaft* 114 (2): 30–32.

Dewey, John. [1927] 1988. *The Public and Its Problems*. Athens: Ohio University Press.

Dewey, John. 1929. *The Quest for Certainty: A Study of the Relation of Knowledge and Action*. London: George Allen & Unwin.

Dietz, Hella. 2004. Unbeabsichtigte Folgen: Hauptbegriff der Soziologie oder verzichtbares Konzept. *Zeitschrift für Soziologie* 33 (1): 48–61.

Dodgson, Mark. 1993. Learning, Trust, and Technological Collaboration. *Human Relations* 46 (1): 77–95.

Dorf, Michael C., and Charles F. Sabel. 1998. A Constitution of Democratic Experimentalism. *Columbia Law Review* 98 (2): 267–473.

Dovers, Stephen, Tony W. Norton, and John W. Handmer. 1996. Uncertainty, Ecology, Sustainability and Policy. *Biodiversity and Conservation* 5 (10): 1143–1167.

Drenthen, Martin, Jozef Keulartz, and James Proctor, eds. 2009. *New Visions of Nature: Complexity and Authenticity*. Berlin: Springer.

Drucker, Peter F. 1973. *Management: Tasks, Responsibilities, Practices.* New York: Harper & Row.

Ducatel, Ken. 2000. Information Technologies in Non-knowledge Services: Innovations at the Margin? In *Innovation Systems in the Service Economy,* ed. J. Stanley Metcalfe and Ian Miles, 221–245. Boston: Kluwer.

Duncan, Ian J. 1999. A Community That Accepts Risk Should Be Rewarded. *Risk Decision and Policy* 4 (3): 191–199.

Durkheim, Emile. [1893] 1933. *The Division of Labor in Society.* Glencoe, IL: Free Press.

Ebeling, Johannes. 2002. Wie verändert Gentechnik die Gesellschaft? In *Designing Human Beings: Die politische Dimension der Biotechnologie,* ed. Klaus Segbers and Dirk Lullies, 47–54. Berlin: Osteuropa-Institut der Freien Universität Berlin.

Eden, Sally, and Sylvia Tunstall. 2006. Ecological versus Social Restoration? How Urban River Restoration Challenges but Also Fails to Challenge the Science-Policy Nexus in the United Kingdom. *Environment and Planning C* 24 (5): 661–680.

Edgerton, David. 2004. "The Linear Model" Did Not Exist: Reflections on the History and Historiography of Science and Research in Industry in the Twentieth Century. In *The Science-Industry Nexus: History, Policy, Implications,* ed. Karl Grandin, Nina Wormbs, and Sven Widmalm, 31–58. Sagamore Beach, MA: Science History Publications.

Eglash, Ron, Jennifer L. Croissant, Giovanni Di Chiro, and Rayvan Fouché, eds. 2004. *Appropriating Technology: Vernacular Science and Social Power.* Minneapolis: University of Minnesota Press.

Elvers, Horst-Dietrich, Matthias Gross, and Harald Heinrichs. 2008. The Diversity of Environmental Justice: Towards a European Approach. *European Societies* 10 (5): 835–856.

Enserink, Martin. 2008. Malaria: Vaccine Comes Another Step Closer. *Science* 322 (5908): 1622–1623.

Etzioni, Amitai. 1967. Mixed Scanning: A Third Approach to Decision Making. *Public Administration Review* 27 (5): 385–392.

Etzkowitz, Henry. 2008. *The Triple Helix: University-Industry-Government Innovation in Action.* London: Routledge.

Faber, Malte, Reiner Manstetten, and John L. R. Proops. 1992. Humankind and the Environment: An Anatomy of Surprise and Ignorance. *Environmental Values* 1 (3): 217–241.

Faber, Malte, and John L. R. Proops. [1990] 1998. *Evolution, Time, Production and the Environment.* Berlin: Springer.

Feldman, Roberta M. 2007. *Post-Occupancy Evaluation of Restored Natural Areas in Chicago's Lincoln Park. Chicago: City Design Center and College of Architecture and the Arts*. Chicago: University of Illinois.

Felt, Ulrike, Brian Wynne, Michel Callon, Maria E. Gonçalves, Sheila Jasanoff, Maria Jepsen, Pierre-Benoît Joly, et al. 2007. *Taking European Knowledge Society Seriously*. Luxembourg: Office for Official Publications of the European Communities.

Fine, Gary Alan. 2006. The Chaining of Social Problems: Solutions and Unintended Consequences in the Age of Betrayal. *Social Problems* 53 (1): 3–17.

Fine, Gary Alan. 2007. *Authors of the Storm: Meteorologists and the Culture of Prediction*. Chicago: University of Chicago Press.

Fleck, Ludwik. [1935] 1979. *Genesis and Development of a Scientific Fact*. Chicago: University of Chicago Press.

Floricel, Serghei, and Roger Miller. 2001. Strategizing for Anticipated Risks and Turbulence in Large-Scale Engineering Projects. *International Journal of Project Management* 19 (8): 445–455.

Folke, Carl. 2006. Resilience: The Emergence of a Perspective for Social-Ecological Systems Analyses. *Global Environmental Change* 16 (3): 253–267.

Foray, Dominique, and Michael Gibbons. 1996. Discovery in the Context of Application. *Technological Forecasting and Social Change* 53 (3): 263–277.

Forester, John. 1984. Bounded Rationality and the Politics of Muddling Through. *Public Administration Review* 44 (1): 23–31.

Forman, Paul. 2007. The Primacy of Science in Modernity, of Technology in Postmodernity, and of Ideology in the History of Technology. *History and Technology* 23 (1–2): 1–152.

Fowler, Martin. 1999. *Refactoring: Improving the Design of Existing Code*. New York: Addison-Wesley.

Freudenburg, William R., Robert Gramling, and Debra J. Davidson. 2008. Scientific Certainty Argumentation Methods (SCAMs): Science and the Politics of Doubt. *Sociological Inquiry* 78 (1): 2–38.

Frickel, Scott, Sahra Gibbon, Jeff Howard, Joanna Kempner, Gwen Ottinger, and David Hess. 2010. Undone Science: Charting Social Movement and Civil Society Challenges to Research Agenda Setting. *Science, Technology and Human Values* 35.

Frickel, Scott. 2008. On Missing New Orleans: Lost Knowledge and Knowledge Gaps in an Urban Hazardscape. *Environmental History* 13 (4):41–65.

Frickel, Scott, and Kelly Moore, eds. 2006. *The New Political Sociology of Science: Institutions, Networks, and Power*. Madison: University of Wisconsin Press.

Friederici, Peter. 2006. *Nature's Restoration: People and Places on the Front Lines of Conservation.* Covelo, CA: Island Press.

Fritz, Wolfgang, and Friedrich-Carl Benthaus. 2000. Application of New Techniques to Create Post-Mining Landscape Suitable for Future Generations. *Braunkohle – Surface Mining* 52 (3): 261–265.

Frodeman, Robert. 2005. Acid Mine Drainage. In *Encyclopedia of Science, Technology, and Ethics*, vol. 1, ed. Carl Mitcham., 11–12. Detroit: Thomson Gale.

Fuller, Steve. 2000. *The Governance of Science: Ideology and the Future of the Open Society.* Buckingham, UK: Open University Press.

Fuller, Steve. 2001. A Critical Guide to Knowledge Society Newspeak: Or, How Not to Take the Great Leap Backward. *Current Sociology* 49 (4): 177–201.

Funtowicz, Silvio O., and Jerome R. Ravetz. 1990. *Uncertainty and Quality in Science for Policy.* Dordrecht: Kluwer.

Funtowicz, Silvio O., and Jerome R. Ravetz. 1993. Science for the Post-Normal Age. *Futures* 25 (7): 739–755.

Furedi, Frank. 2006. *Culture of Fear Revisited.* London: Continuum.

Furnweger, Kate. 1997. The Magic Hedge at Montrose Point: Plans for Development. *Compass* 11 (7–8): 1–4.

Galison, Peter. 1997. *Image and Logic: A Material Culture of Microphysics.* Chicago: University of Chicago Press.

Gannon, Frank. 2008. The End of Optimism? *EMBO Reports* 9 (2): 111.

Geels, Frank, and J. Jasper Deuten. 2006. Local and Global Dynamics in Technological Development: A Socio-cognitive Perspective on Knowledge Flows and Lessons from Reinforced Concrete. *Science & Public Policy* 33 (4): 265–275.

Geissler-Strobel, Sabine, Jens Bugner, Reinart Feldmann, Klaus Günther, Josefine Gras, Frank Herbst, and Kerstin Seluga. 1998. Bergbaufolgelandschaften in Ostdeutschland: Durch Sanierung bedrohte Sekundärlebensräume. *Naturschutz und Landschaftsplanung* 30 (4): 106–114.

Gerke, Solvay, and Hans-Dieter Evers. 2006. Globalizing Local Knowledge: Social Science Research on Southeast Asia, 1970–2000. *Sojourn: Journal of Social Issues in Southeast Asia* 21 (1): 1–21.

Giardina, Christian P., Creighton M. Litton, Jarrod M. Thaxton, Susan Cordell, Lisa J. Hadway, and Darren R. Sandquist. 2007. Science Driven Restoration: A Candle in a Demon Haunted World. Response to Cabin (2007). *Restoration Ecology* 15 (2): 171–176.

Gibbons, Michael, Camille Limoges, Helga Nowotny, Simon Schwartzman, Peter Scott, and Martin Trow. 1994. *The New Production of Knowledge: The Dynamics of Science and Research in Contemporary Societies*. London: Sage.

Giddens, Anthony. 1990. *The Consequences of Modernity*. Stanford: Stanford University Press.

Giddens, Anthony. 2000. *Runaway World: How Globalization is Reshaping Our Lives*. New York: Routledge.

Gieryn, Thomas F. 1999. *Cultural Boundaries of Science: Credibility on the Line*. Chicago: University of Chicago Press.

Gieryn, Thomas F. 2006. City as Truth-Spot: Laboratories and Field-Sites in Urban Studies. *Social Studies of Science* 36 (1): 5–38.

Gigerenzer, Gerd, and Reinhard Selten, eds. 2001. *Bounded Rationality: The Adaptive Toolbox*. Cambridge, MA: MIT Press.

Gobster, Paul H. 1997. The Chicago Wilderness and Its Critics III: The Other Side. A Survey of the Arguments. *Restoration and Management Notes* 15 (1): 33–38.

Gobster, Paul H. 2001. Visions of Nature: Conflict and Compatibility in Urban Park Restoration. *Landscape and Urban Planning* 56 (1–2): 35–51.

Gobster, Paul H. 2004. The Urban Restoration Experience. Paper presented at the San Francisco Natural History Lecture Series, Randall Museum, San Francisco, CA, April 22.

Gobster, Paul H. 2007. Urban Park Restoration and the "Museumification" of Nature. *Nature and Culture* 2 (2): 95–114.

Gobster, Paul H., and Susan C. Barro. 2000. Negotiating Nature: Making Restoration Happen in a Urban Park Context. In *Restoring Nature: Perspectives from the Social Sciences and Humanities*, ed. Paul H. Gobster and Bruce Hull, 185–207. Covelo, CA: Island Press.

Godin, Benoît. 1998. Writing Performative History: The New New Atlantis? *Social Studies of Science* 28 (3): 465–483.

Godin, Benoît. 2006. The Linear Model of Innovation: The Historical Construction of an Analytical Framework. *Science, Technology and Human Values* 31 (6): 639–667.

Gooding, David C. 1993. What Is Experimental about Thought Experiments? In *Proceedings of the 1992 Biennial Meeting of the Philosophy of Science Association*, ed. David Hull, Mickey Forbers, and Kathleen Okruhlik, vol. 2, 280–290. East Lansing, MI: Philosophy of Science Association.

Grabher, Gernot. 1993. The Weakness of Strong Ties: The Lock-in of Regional Development in the Ruhr Area. In *The Embedded Firm*, ed. Gernot Grabher, 255–277. London: Routledge.

Greenberg, Joel. 2002. *A Natural History of the Chicago Region*. Chicago: University of Chicago Press.

Greenwood, Davydd, and Morten Levin. 1998. *Introduction to Action Research: Social Research for Social Change*. Thousand Oaks, CA: Sage.

Greenwood, Ernest. [1945] 1976. *Experimental Sociology: A Study in Method*. New York: Octagon Books.

Greshoff, Rainer, Georg Kneer, and Uwe Schimank, eds. 2003. *Die Transintentionalität des Sozialen. Eine vergleichende Betrachtung klassischer und moderner Sozialtheorien*. Opladen: Westdeutscher Verlag.

Gross, Matthias. 2001. *Die Natur der Gesellschaft: Eine Geschichte der Umweltsoziologie*. Weinheim: Juventa Verlag.

Gross, Matthias. 2002. New Natures and Old Science: Hands-on Practice and Academic Research in Ecological Restoration. *Science Studies* 15 (2): 17–35.

Gross, Matthias. 2003a. *Inventing Nature: Ecological Restoration by Public Experiments*. Lanham, MD: Lexington Books.

Gross, Matthias. 2003b. Sociologists of the Unexpected: Edward A. Ross and Georg Simmel on the Unintended Consequences of Modernity. *American Sociologist* 34 (4): 40–58.

Gross, Matthias. 2004a. Human Geography and Ecological Sociology: The Unfolding of a Human Ecology, 1890 to 1930—and Beyond. *Social Science History* 28 (4): 575–605.

Gross, Matthias. 2004b. A Unique American Sociology or an Americanized European Sociology? Some Thoughts on Georg Simmel's Influence on Edward A. Ross. *Simmel Studies* 14 (2): 195–222.

Gross, Matthias. 2006. Community by Experiment: Recursive Practice in Landscape Design and Ecological Restoration. In *Community and Ecology: Dynamics of Place, Sustainability, and Politics*, ed. Aaron M. McCright and Terry N. Clark, 47–66. Oxford: Elsevier.

Gross, Matthias. 2007a. Restoration and the Origins of Ecology. *Restoration Ecology* 15 (3): 375–376.

Gross, Matthias. 2007b. The Unknown in Process: Dynamic Connections of Ignorance, Non-Knowledge and Related Concepts. *Current Sociology* 55 (5): 742–759.

Gross, Matthias. 2008. Population Decline and the New Nature: Towards Experimental "Refactoring" in Landscape Development of Post-industrial Regions. *Futures* 40 (5): 451–459.

Gross, Matthias. 2009. Collaborative Experiments: Jane Addams, Hull-House, and Experimental Social Work. *Social Sciences Information. Information sur les sciences sociales* 48 (1): 81–95.

Gross, Matthias, Holger Hoffmann-Riem, and Wolfgang Krohn. 2003. Realexperimente: Robustheit und Dynamik ökologischer Gestaltungen in der Wissensgesellschaft. *Soziale Welt* 54 (3): 241–258.

Gross, Matthias, and Wolfgang Krohn. 2005. Society as Experiment: Sociological Foundations for a Self-Experimental Society. *History of the Human Sciences* 18 (2): 63–86.

Groves, Christopher. 2009. Nanotechnology, Contingency, and Finitude. *NanoEthics* 3 (1): 1–16.

Gubbins, Claire, and Sarah MacCurtain. 2008. Understanding the Dynamics of Collective Learning: The Role of Trust and Social Capital. *Advances in Developing Human Resources* 10 (4): 578–599.

Gunderson, Lance L. 2003. Adaptive Dancing: Interactions between Social Resilience and Ecological Crisis. In *Navigating Social-Ecological Systems*, ed. Fikret Berkes, Johan Colding, and Carl Folke, 33–52. Cambridge: Cambridge University Press.

Gunderson, Lance L., and Crawford S. Holling, eds. 2002. *Panarchy: Understanding Transformations in Systems of Humans and Nature*. Washington, DC: Island Press.

Gunkel, Günter. 1996. *Renaturierung kleiner Fließgewässer*. Jena: Fischer.

Haase, Dagmar. 2008. Urban Ecology of Shrinking Cities: An Unrecognized Opportunity? *Nature and Culture* 3 (1): 1–8.

Habermas, Jürgen. 1984. *The Theory of Communicative Action*. (2 vols.). Boston: Beacon Press.

Hacking, Ian. 1983. *Representing and Intervening: Introductory Topics in the Philosophy of Natural Science*. Cambridge: Cambridge University Press.

Halfon, Saul. 2007. Science and the Precautionary Principle. In *The Blackwell Encyclopedia of Sociology*, ed. George Ritzer, 4082–4083. Williston, VT: Blackwell.

Halle, Stefan. 2007. Present State and Future Perspectives of Restoration Ecology. *Restoration Ecology* 15 (2): 304–306.

Harding, Sandra. 2008. *Sciences from Below: Feminisms, Postcolonialities, and Modernities*. Durham, NC: Duke University Press.

Harloe, Michael, and Beth Perry. 2004. Universities, Localities and Regional Development: The Emergence of the "Mode 2" University? *International Journal of Urban and Regional Research* 28 (1): 212–223.

Harris, James A., Richard J. Hobbs, Eric Higgs, and James Aronson. 2006. Ecological Restoration and Global Climate Change. *Restoration Ecology* 14 (2): 170–176.

Hassink, Robert. 2005. How to Unlock Regional Economies from Path Dependency? From Learning Region to Learning Cluster. *European Planning Studies* 13 (4): 521–535.

Hayek, Friedrich A. 1945. The Use of Knowledge in Society. *American Economic Review* 35 (4): 519–530.

Hedberg, Bo. 1981. How Organizations Learn and Unlearn. In *Handbook of Organizational Design*. vol. 1, ed. Paul C. Nystrom and William H. Starbuck, 3–27. Oxford: Oxford University Press.

Helford, Reid M. 2003. *Prairie Politics: Constructing Science, Nature and Community in the Chicago Wilderness*. Dissertation thesis, Department of Sociology, Loyola University Chicago.

Herbold, Ralf. 1995. Technologies as Social Experiments: The Construction and Implementation of a High-Tech Waste Disposal. In *Managing Technology in Society*, ed. Ari Rip, Thomas J. Misa, and Johan Schot, 185–197. London: Pinter.

Herwig, Wolfram. 2001. Stopp Cospuden 90: Auseinandersetzung um einen Tagebau. *Südraum Journal* 12:4–9.

Hess, David J. 2007. *Alternative Pathways in Science and Industry: Activism, Innovation, and the Environment in an Era of Globalization*. Cambridge, MA: MIT Press.

Hess, David J. 2009. The Potentials and Limitations of Civil Society Research: Getting Undone Science Done. *Sociological Inquiry* 79 (3): 306–327.

Hessels, Laurens K., and Harro van Lente. 2008. Re-thinking New Knowledge Production: A Literature Review and a Research Agenda. *Research Policy* 37 (4): 740–760.

Higgs, Eric S. 1994. Expanding the Scope of Restoration Ecology. *Restoration Ecology* 2 (3): 137–146.

Higgs, Eric S. 2003. *Nature by Design: People, Natural Process, and Ecological Restoration*. Cambridge: MIT Press.

Higgs, Eric S. 2005. The Two-Culture Problem: Ecological Restoration and the Integration of Knowledge. *Restoration Ecology* 13 (1): 159–164.

Hinkle, Gisela J. 1992. Habermas, Mead, and Rationality. *Symbolic Interaction* 15 (3): 315–331.

Hobbs, Richard J. 2009. Looking for the Silver Lining: Making the Most of Failure. *Restoration Ecology* 17 (1): 1–3.

Hobbs, Richard J., and James A. Harris. 2001. Restoration Ecology: Repairing the Earth's Ecosystems in the New Millennium. *Restoration Ecology* 9 (2): 239–246.

Hoffmann-Riem, Holger, and Brian Wynne. 2002. In Risk Assessment, One Has to Admit Ignorance. *Nature* 416:123. http://www.ncbi.nlm.nih.gov/entrez/query .fcgi?cmd=Retrieve&db=PubMed&list_uids=11894070&dopt=Abstract.

Holling, Crawford S., ed. 1978. *Adaptive Environmental Assessment and Management.* London: Wiley.

Holling, Crawford S. 1986. The Resilience of Terrestrial Ecosystems: Local Surprise and Global Change. In *Sustainable Development of the Biosphere*, ed. William C. Clark and Robert E. Munn, 292–317. Cambridge: Cambridge University Press.

Hood, Christopher, and Henry Rothstein. 2001. Risk Regulation under Pressure: Problem Solving or Blame Shifting? *Administration and Society* 33 (1): 21–53.

Huber, Joseph. 2004. *New Technologies and Environmental Innovation.* Cheltenham, UK: Elgar.

Huber, Michael. 2008. Fundamental Ignorance in the Regulation of Reactor Safety and Flooding. In *Who Owns Knowledge? Knowledge and the Law*, ed. Nico Stehr and Bernd Weiler, 107–124. New York: Transaction.

Huff, Anne S. 2000. Changes in Organizational Knowledge Production. *Academy of Management Review* 25 (2): 288–293.

Hughes, Thomas P. 1987. The Evolution of Large Technological Systems. In *The Social Construction of Technological Systems: New Directions in the Sociology and History of Technology*, ed. Wiebe E. Bijker, Thomas P. Hughes, and Trevor Pinch, 51–82. Cambridge, MA: MIT Press.

Hüser, Christian. 2006. Robustness: A Challenge also for the Twenty-first Century. Department of Ecological Modeling, Helmholtz Centre for Environmental Research, Discussion Paper 2/2006, Leipzig, Germany.

Iles, Alastair. 2007. Identifying Environmental Health Risks in Consumer Products: Non-governmental Organizations and Civic Epistemologies. *Public Understanding of Science* 16 (4): 371–391.

Ingram, Mrill. 2009. You Don't Have to Be a Scientist to Do Science. *Ecological Restoration* 27 (1): 1–2.

Ito, Junitsu, Anil Ghosh, Luciano A. Moreira, Ernst A. Wimmer, and Marcelo Jacobs-Lorena. 2002. Transgenic Anopheline Mosquitoes Impaired in Transmission of a Malaria Parasite. *Nature* 417: 452–455.

Jäger diskutieren über Rückkehr der Wölfe. 2007. *Berliner Morgenpost*. May 6.

Janssen, Marco A., and Pim Martens. 1997. Modeling Malaria as a Complex Adaptive System. *Artificial Life* 3 (3): 213–236.

Japp, Klaus P. 2000. Distinguishing Non-Knowledge. *Canadian Journal of Sociology* 25 (2): 225–238.

Jasanoff, Sheila. 1990. *The Fifth Branch: Science Advisers as Policymakers*. Cambridge, MA: Harvard University Press.

Jasanoff, Sheila. 2005. *Designs on Nature: Science and Democracy in Europe and the United States*. Princeton, NJ: Princeton University Press.

Jaspers, Karl. 1951. *Way to Wisdom: An Introduction to Philosophy*. New Haven, CT: Yale University Press.

Jaworski, Gary D. 1997. *Georg Simmel and the American Prospect*. Albany: State University of New York Press.

Jokisch, Rodigo. 1981. Die nichtintentionalen Effekte menschlicher Handlungen. Ein klassisches soziologisches Problem. *Kölner Zeitschrift für Soziologie und Sozialpsychologie* 33 (3): 547–575.

Jones, Dawn S. 2007. "Fearing the Worst, Hoping for the Best": The Discursive Construction of Risk in Pregnancy. In *Risks, Identities and the Everyday*, ed. Julie Scott Jones and Jayne Raisborough, 37–50. Aldershot, UK: Ashgate.

Jordan, William R., III. 1990. Making an Urban Wilderness: Reflections on the First Fifty Years of the University of Wisconsin Arboretum. In *Green Cities: Ecologically Sound Approaches to Urban Space*, ed. David Gordon, 67–80. Montreal: Black Rose Books.

Jordan, William R., III. 1994. Sunflower Forest: Ecological Restoration as the Basis for a New Environmental Paradigm. In *Beyond Preservation: Restoring and Inventing Landscapes*, ed. A. Dwight Baldwin, Judith Luce, and Carl Pletsch, 17–34. Minneapolis: University of Minnesota Press.

Jordan, William R., III. 2003. *The Sunflower Forest: Ecological Restoration and the New Communion with Nature*. Berkeley: University of California Press.

Jordan, William R., III. 2006. Ecological Restoration: Carving a Niche for Humans in the Classic Landscape. *Nature and Culture* 1 (1): 22–35.

Jordan, William R., III, Michael E. Gilpin, and John D. Aber, eds. 1987. *Restoration Ecology: A Synthetic Approach to Ecological Research*. Cambridge: Cambridge University Press.

Kabisch, Sigrun. 2004. Revitalisation Chances for Communities in Post-Mining Landscapes. *Peckiana* 3: 87–99.

Kay, Neil M. 1984. *The Emergent Firm: Knowledge, Ignorance and Surprise in Economic Organisation.* London: Macmillan.

Keen, Meg, Valerie A. Brown, and Rob Dyball, eds. 2005. *Social Learning in Environmental Management: Towards a Sustainable Future.* London: Earthscan.

Kerwin, Ann. 1993. None Too Solid: Medical Ignorance. *Knowledge: Creation, Diffusion, Utilization* 15 (2): 166–185.

Kim, Linsu. 1998. Crisis Construction and Organizational Learning: Capability Building in Catching-up at Hyundai Motor. *Organization Science* 9 (4): 506–521.

Klapper, Helmut, and Martin Schultze. 1995. Geogenically Acidified Mining Lakes: Living Conditions and Possibilities of Restoration. *Internationale Revue der Gesamten Hydrobiologie* 80 (4): 639–653.

Kleinman, Daniel L. 2005. *Science and Technology in Society: From Biotechnology to the Internet.* Malden, MA: Blackwell.

Kluth, Gesa, and Ilka Reinhardt. 2005. *Mit Wölfen leben: Informationen für Jäger, Förster und Tierhalter.* Rietschen: Kontaktbüro Wolfsregion Lausitz.

Knight, Frank H. 1921. *Risk, Uncertainty and Profit.* Boston: Houghton Mifflin.

Knorr Cetina, Karin. 1999. *Epistemic Cultures: How the Sciences Make Knowledge.* Cambridge, MA: Harvard University Press.

Knorr Cetina, Karin. 2007. Culture in Global Knowledge Societies: Knowledge Cultures and Epistemic Cultures. *Interdisciplinary Science Reviews* 32 (4): 361–375.

Köck, Wolfgang, Jana Bovet, Thomas Gawron, Ekkehard Hofmann, and Stefan Möckel. 2006. *Effektivierung des raumbezogenen Planungsrechtes am Beispiel der Flächeninanspruchnahme. Final Report for the Umweltbundesamt (UBA).* Leipzig: UFZ Leipzig.

Kowarik, Ingo, and Stefan Körner, eds. 2005. *Wild Urban Woodlands: New Perspectives for Urban Forestry.* Berlin: Springer.

Kricher, John. 2009. *The Balance of Nature: Ecology's Enduring Myth.* Princeton, NJ: Princeton University Press.

Krohn, Wolfgang. 1977. Die Neue Wissenschaft der Renaissance. In *Experimentelle Philosophie. Ursprünge autonomer Wissenschaftsentwicklung*, ed. Gernot Böhme, Wolfgang van den Daele, and Wolfgang Krohn, 13–128. Frankfurt am Main: Suhrkamp.

Krohn, Wolfgang. 2001. Knowledge Societies. In *International Encyclopedia of the Social and Behavioral Sciences*, ed. Neil J. Smelser and Paul B. Baltes, 8139–8143. Oxford: Blackwell.

Krohn, Wolfgang. 2007. Nature, Technology, and the Acknowledgment of Waste. *Nature and Culture* 2 (2): 139–160.

Krohn, Wolfgang, and Johannes Weyer. 1994. Society as a Laboratory: The Social Risks of Experimental Research. *Science and Public Policy* 21 (3): 173–183.

Krüger, Bernd, Andreas Kadler, and Michael Fischer. 2002. The Creation of Post-Mining Landscapes of Lignite Mining in the New Federal States. *Surface Mining* 54 (2): 161–169.

Krummsdorf, Albrecht, and Gerhard Krümmer, eds. 1981. *Landschaft vom Reißbrett: Die Zukunft unserer Kippen, Halden und Restlöcher*. Leipzig: Urania Verlag.

Kübler, Hans-Dieter. 2005. *Mythos Wissensgesellschaft: Gesellschaftlicher Wandel zwischen Information, Medien und Wissen*. Wiesbaden: VS Verlag für Sozialwissenschaften.

Küffer, Christoph. 2006. Integrative Ecological Research: Case-Specific Validation of Ecological Knowledge for Environmental Problem Solving. *Gaia* 15 (2): 115–120.

Kuhlicke, Christian, and Sylvia Kruse. 2009. Nichtwissen und Resilienz in der lokalen Klimaanpassung. *Gaia* 18 (3): 247–254.

Küppers, Günter. 1999. Coping with Uncertainty: New Forms of Knowledge Production. *AI and Society* 13 (1–2): 52–62.

Küppers, Günter. 2002. Complexity, Self-Organisation and Innovation Networks: A New Theoretical Approach. In *Innovation Networks: Theory and Practice*, ed. Andreas Pyka and Günter Küppers., 22–52. Cheltenham, UK: Elgar.

Kwiatkowski, Dominic P. 2005. How Malaria Has Affected the Human Genome and What Human Genetics Can Teach Us about Malaria. *American Journal of Human Genetics* 77 (2): 171–192.

Lahsen, Myanna. 2005. Seductive Simulations? Uncertainty Distribution around Climate Models. *Social Studies of Science* 35 (6): 895–922.

Lane, Robert E. 1966. The Decline of Politics and Ideology in a Knowledgeable Society. *American Sociological Review* 31 (5): 649–662.

Langhorne, Jean, Francis M. Ndungu, Anne-Marit Sponaas, and Kevin Marsh. 2008. Immunity to Malaria: More Questions than Answers. *Nature Immunology* 9 (7): 725–732.

Lash, Scott. 1999. *Another Modernity, a Different Rationality*. Oxford: Blackwell.

Latour, Bruno. 1999. *Pandora's Hope: Essays on the Reality of Science Studies*. Cambridge, MA: Harvard University Press.

Latour, Bruno. 2004. *Politics of Nature: How to Bring the Sciences Into Democracy*. Cambridge, MA: Harvard University Press.

Law, John, and Annemarie Mol, eds. 2002. *Complexities: Social Studies of Knowledge Practices*. Durham, NC: Duke University Press.

Lee, Kai N. 1993. *Compass and Gyroscope: Integrating Science and Politics for the Environment*. Washington, DC: Island Press.

Lee, Raymond L.M. 2008. In Search of Second Modernity: Reinterpreting Reflexive Modernization in the Context of Multiple Modernities. *Social Sciences Information. Information sur les sciences sociales* 47 (1): 55–69.

Lemert, Charles. 1993. Modernity's Classical Age: 1848–1919. In *Social Theory: The Multicultural and Classic Readings*, ed. Charles Lemert, 25–33. Boulder: Westview Press.

Lemov, Rebecca. 2005. *World as Laboratory: Experiments with Mice, Mazes and Men*. New York: Hill and Wang.

Leopold, Aldo. [1941] 1991. Wilderness as a Land Laboratory. In *The River of the Mother of God and Other Essays by Aldo Leopold*, ed. Susan L. Flader and J. Baird Callicot, 287–289. Madison: University of Wisconsin Press.

Levidow, Les, and Susan Carr. 2007. GM Crops on Trial: Technological Development as a Real-World Experiment. *Futures* 39 (4): 408–431.

Levine, Donald N. 2000. On the Critique of "Utilitarian" Theories of Action: Newly Identified Convergences among Simmel, Weber, and Parsons. *Theory, Culture and Society* 17 (1): 63–78.

Lezaun, Javier, and Yuval Millo. 2004. Testing Times: What Is the Purpose of Regulatory Experiments? *Risk and Regulation* 7: 8–9.

Light, Andrew. 2003. Tolkien's Green Time: Environmental Themes in *The Lord of the Rings*. In *The Lord of the Rings and Philosophy*, ed. Gregory Bassham and Eric Bronson, 150–163. Chicago: Open Court.

Light, Andrew. 2006. Ecological Citizenship: The Democratic Promise of Restoration. In *The Humane Metropolis: People and Nature in the 21st Century*, ed. Rutherford H. Platt, 169–182. Amherst: University of Massachusetts Press.

Lightman, Bernard. 1987. *The Origins of Agnosticism: Victorian Unbelief and the Limits of Knowledge*. Baltimore, MD: Johns Hopkins University Press.

Lindblom, Charles E. 1959. The Science of "Muddling Through." *Public Administration Review* 19 (2): 79–88.

Linke, Sabine, and Lutz Schiffer. 2002. Development Prospects for the Post-Mining Landscape in Central Germany. In *Remediation of Abandoned Surface Coal Mining Sites*, ed. Alena Murdoch, Ulrich Stottmeister, Christopher Kennedy, and Helmut Klapper, 111–149. Berlin: Springer.

LMBV. 2003. *Sanierungsbericht 2003*. Berlin: Unpublished Report of the Lusatian and Central German Mining Administration Company.

Lopata, Helena Z. 1976. Expertization of Everyone and the Revolt of the Client. *Sociological Quarterly* 17 (4): 435–447.

Luhmann, Niklas. 1990. *Die Wissenschaft der Gesellschaft*. Frankfurt am Main: Suhrkamp.

Luhmann, Niklas. 1992. *Beobachtungen der Moderne*. Opladen: Westdeutscher Verlag.

Luhmann, Niklas. 1993. *Risk: A Sociological Theory*. New York: Aldine de Gruyter.

Luhmann, Niklas. 2000. *The Reality of Mass Media*. Stanford: Stanford University Press.

Luhmann, Niklas. 2005. *Einführung in die Theorie der Gesellschaft*. Heidelberg: Carl-Auer-Verlag.

Luke, Timothy W. 2005. The Death of Environmentalism or the Advent of Public Ecology? *Organization and Environment* 18 (4): 489–494.

Machlup, Fritz. 1962. *The Production and Distribution of Knowledge in the United States*. Princeton, NJ: Princeton University Press.

Machlup, Fritz. 1980. *Knowledge and Knowledge Production*. Vol. 1, Knowledge: Its Creation, Distribution, and Economic Significance. Princeton, NJ: Princeton University Press.

March, James G., and Herbert A. Simon. 1958. *Organizations*. New York: Wiley.

Markham, William T. 2008. *Environment Organizations in Modern Germany: Hardy Survivors in the Twentieth Century and Beyond*. New York: Berghahn Books.

Marmorek, David. 2004. Adaptive Management: Theory and Practice. Paper presented at the workshop on Real-World Experiments: Combining Ecological Research and Design, Bielefeld, September.

Martin, Ben R., and John Irvine. 1989. *Research Foresight: Priority-Setting in Science*. London: Pinter.

Martin, Frank E. 2005. Negotiating Nature. *Landscape Architecture* 95 (1): 112–115.

May, Stefan. 2003. Nebenfolgen: Veränderungen im Recht durch Nichtwissen in der Biomedizin. In *Wissenschaft in der Wissensgesellschaft*, ed. Stefan Böschen and Ingo Schulz-Schaeffer, 236–249. Wiesbaden: Westdeutscher Verlag.

McAllister, James W. 2004. Thought Experiments and the Belief in Phenomena. *Philosophy of Science* 71 (5): 1164–1175.

McDaniel, Reuben R., Jr., Michelle E. Jordan, and Brigitte F. Fleeman. 2003. Surprise, Surprise, Surprise! A Complexity Science View of the Unexpected. *Health Care Management Review* 28 (3): 266–278.

McElhinney, Stephen. 2005. Exposing the Interests: Decoding the Promise of the Global Knowledge Society. *New Media and Society* 7 (6): 748–769.

McNamee, Thomas. 1997. *The Return of the Wolf to Yellowstone*. New York: Holt.

Mead, George Herbert. 1934. *Mind, Self, and Society*. Chicago: University of Chicago Press.

Meeker, Barbara F., and Robert K. Leik. 1995. Experimentation in Sociological Social Psychology. In *Sociological Perspectives on Social Psychology*, ed. Karen S. Cook, Gary Alan Fine, and James S. House, 629–649. Boston: Allyn and Bacon.

Merton, Robert K. 1936. The Unanticipated Consequences of Purposive Social Action. *American Sociological Review* 1 (6): 894–904.

Merton, Robert K. [1949] 1968. *Social Theory and Social Structure*. New York: Free Press.

Merton, Robert K. 1987. Three Fragments from a Sociologist's Notebook: Establishing the Phenomenon, Specified Ignorance, and Strategic Research Materials. *Annual Review of Sociology* 13:1–28.

Merton, Robert K. 1995. The Thomas Theorem and the Matthew Effect. *Social Forces* 74 (2): 379–424.

Merton, Robert K., and Eleanor Barber. 2004. *The Travels and Adventures of Serendipity: A Study in Sociological Semantics and the Sociology of Science*. Princeton: Princeton University Press.

Milgram, Stanley. 1974. *Obedience to Authority: An Experimental View*. London: Tavistock.

Miller, Clark A. 2005. New Civic Epistemologies of Quantification: Making Sense of Indicators of Local and Global Sustainability. *Science, Technology and Human Values* 30 (3): 403–432.

Miller, Peter, and Ted O'Leary. 1994. The Factory as Laboratory. *Science in Context* 7 (3): 469–496.

Miller, William Ian. 2000. *The Mystery of Courage*. Cambridge, MA: Harvard University Press.

Minkler, Meredith, ed. 2004. *Community Organizing and Community Building for Health*. New Brunswick, NJ: Rutgers University Press.

Mittelstrass, Jürgen. 1996. Nichtwissen: Preis des Wissens? *Schweizerische Technische Zeitschrift* 93 (6): 32–35.

Mol, Arthur P.J. 2001. *Globalization and Environmental Reform: The Ecological Modernization of the Global Economy.* Cambridge, MA: MIT Press.

Mol, Arthur P.J. 2008. *Environmental Reform in the Information Age: The Contours of Informational Governance.* Cambridge: Cambridge University Press.

Montrie, Chad. 2003. *To Save the Land and People: A History of Opposition to Surface Coal Mining in Appalachia.* Chapell Hill, NC: University of North Carolina Press.

Moore, Kelly. 2006. Powered by the People: Scientific Authority in Participatory Science. In *The New Political Sociology of Science: Institutions, Networks, and Power,* ed. Scott Frickel and Kelly Moore, 299–323. Madison: University of Wisconsin Press.

Moore, Kelly. 2008. *Disrupting Science: American Scientists and the Politics of the Military, 1945–1975.* Princeton, NJ: Princeton University Press.

Moore, Wilbert E., and Melvin M. Tumin. 1949. Some Social Functions of Ignorance. *American Sociological Review* 14 (6): 787–795.

Moskovits, Debra K., Carol J. Fialkowski, Gregory M. Mueller, and Timothy A. Sullivan. 2002. Chicago Wilderness: A New Force in Urban Conservation. *Annals of the Missouri Botanical Garden* 89 (2): 153–163.

Murdoch, Alena, Ulrich Stottmeister, Christopher Kennedy, and Helmut Klapper, eds. 2002. *Remediation of Abandoned Surface Coal Mining Sites.* Berlin: Springer.

Murphy, Raymond. 2009. *Leadership in Disaster: Learning for a Future with Global Climate Change.* Montreal, Canada: McGill-Queen's University Press.

Myers, Nancy J., and Carolyn Raffensperger. 2005. Green Systems. In *Precautionary Tools for Reshaping Environmental Policy,* ed. Nancy J. Myers and Carolyn Raffensperger, 81–90. Cambridge, MA: MIT Press.

Myers, Norman. 1995. Environmental Unknowns. *Science* 269 (5222): 358–360.

Nichols, Lawrence T. 2001. Parsons and Simmel at Harvard: Scientific Paradigms and Organizational Culture. In *Talcott Parsons Today,* ed. A. Javier Treviño, 1–28. Lanham, MD: Rowman & Littlefield.

Nowotny, Helga. 2005. The Increase of Complexity and its Reduction: Emergent Interfaces between the Natural Sciences, Humanities and Social Sciences. *Theory, Culture and Society* 22 (5): 15–31.

Nowotny, Helga, Peter Scott, and Michael Gibbons. 2001. *Re-Thinking Science: Knowledge and the Public in an Age of Uncertainty.* Oxford: Polity Press.

Nuissl, Henning, and Dieter Rink. 2005. The "Production" of Urban Sprawl in Eastern Germany as a Phenomenon of Post-socialist Transformation. *Cities (London, England)* 22 (2): 123–134.

O'Brien, William E. 2006. Exotic Invasions, Nativism, and Ecological Restoration: On the Persistence of a Contentious Debate. *Ethics Place and Environment* 9 (1): 63–77.

O'Brien, William E., and Jennifer A. McIvor. 2007. Is There Anything "Good" about Everglades Restoration? *Environments* 35 (1): 1–19.

Oreskes, Naomi. 2004. Science and Public Policy: What's Proof Got to Do with It? *Environmental Science and Policy* 7 (5): 369–383.

O'Riordan, Timothy, and James Cameron. 1995. The History and Contemporary Significance of the Precautionary Principle. In *Interpreting the Precautionary Principle*, ed. Timothy O'Riordan and James Cameron, 12–29. London: Earthscan.

Oswalt, Philipp, Klaus Overmeyer, and Walter Prigge. 2002. Experiment und Utopie im Stadtumbau Ostdeutschlands. *Berliner Debatte Initial* 13 (2): 57–63.

Packard, Stephen. 1988. Just a Few Oddball Species: Restoration and the Rediscovery of the Tallgrass Savanna. *Restoration and Management Notes* 6 (1): 13–20.

Park, Robert E. 1921. Sociology and the Social Sciences: The Group Concept and Social Research. *American Journal of Sociology* 27 (2): 169–183.

Park, Robert E. 1929. The City as a Social Laboratory. In *Chicago: An Experiment in Social Science Research*, ed. Thomas V. Smith and Leslie D. White, 1–19. Chicago: University of Chicago Press.

Park, Robert E. 1940. News as a Form of Knowledge: A Chapter in the Sociology of Knowledge. *American Journal of Sociology* 45 (5): 669–686.

Parthey, Heinrich, and Dietrich Wahl. 1966. *Die experimentelle Methode in Natur- und Gesellschaftswissenschaften*. Berlin: Deutscher Verlag der Wissenschaften.

Perin, Constance. 2005. *Shouldering Risks: The Culture of Control in the Nuclear Power Industry*. Princeton, NJ: Princeton University Press.

Perrow, Charles. 1984. *Normal Accidents: Living with High-Risk Technologies*. New York: Basic Books.

Pestre, Dominique. 2000. The Production of Knowledge between Academies and Markets: A Historical Reading of the Book *The New Production of Knowledge*. *Science, Technology & Society* 5 (2): 169–181.

Petroski, Henry. 2006. *Success through Failure: The Paradox of Design*. Princeton, NJ: Princeton University Press.

Pielke, Roger Jr. R. 2005. The Uncertainty Trap. *Prometheus: The Science Policy Blog*. Online at http://sciencepolicy.colorado.edu/prometheus/ (last accessed on July 30, 2008).

Pimm, Stuart L. 1992. *The Balance of Nature? Ecological Issues in the Conservation of Species and Communities*. Chicago: University of Chicago Press.

Pinch, Trevor J. 1981. The Sun-Set: The Presentation of Certainty in Scientific Life. *Social Studies of Science* 11 (1): 131–158.

Plé, Bernhard. 1997. Das Problem der unbeabsichtigten Folgen menschlichen Handelns: Zur Fokussierung eines bleibenden Problems (mit Ansätzen zu einer Typenbildung). *Geschichte und Gegenwart* 16 (3): 179–191.

Pohl, Christian, and Gertrude Hirsch Hadorn. 2007. *Principles for Designing Transdisciplinary Research*. Munich: Oekom.

Polanyi, Michael. 1958. *Personal Knowledge*. London: Routledge and Kegan Paul.

Popitz, Heinrich. 1968. *Über die Präventivwirkung des Nichtwissens*. Tübingen: Mohr.

Portes, Alejandro. 2000. The Hidden Abode: Sociology as Analysis of the Unexpected. *American Sociological Review* 65 (1): 1–18.

Prescott, Lansing M., John P. Harley, and Donald A. Klein. 2005. *Microbiology*. Boston: McGraw-Hill.

Pretty, Jules. 1995. Participatory Learning for Sustainable Agriculture. *World Development* 23 (8): 1247–1263.

Price, Derek J. de Solla. 1961. *Science since Babylon*. New Haven: Yale University Press.

Proctor, Robert N., and Londa Schiebinger, eds. 2008. *Agnotology: The Making and Unmaking of Ignorance*. Stanford: Stanford University Press.

Rauschmayer, Felix, and Heidi Wittmer. 2006. Evaluating Deliberative and Analytical Methods for the Resolution of Environmental Conflicts. *Land Use Policy* 23 (1): 108–122.

Ravetz, Jerome R. 1993. The Sin of Science: Ignorance of Ignorance. *Knowledge: Creation, Diffusion, Utilization* 15 (2): 157–165.

Rayner, Steve. 2000. Prediction and other Approaches to Climate Change Policy. In *Prediction: Science, Decision Making, and the Future of Nature*, ed. Daniel Sarewitz, Roger A. Pielke, and Radford Byerly, 269–296. Washington, DC: Island Press.

Reckwitz, Andreas. 2002. Toward a Theory of Social Practices: A Development in Culturalist Theorizing. *European Journal of Social Theory* 5 (2): 243–263.

Reinhardt, Ilka, and Gesa Kluth. 2007. *Leben mit Wölfen: Leitfaden für den Umgang mit einer konfliktträchtigen Tierart in Deutschland*. Bonn: Bundesamt für Naturschutz (BfN).

Renn, Ortwin. 2008. *Risk Governance: Coping with Uncertainty in a Complex World*. London: Earthscan.

Rheinberger, Hans-Jörg. 1997. *Toward a History of Epistemic Things: Synthesizing Proteins in the Test Tube.* Stanford: Stanford University Press.

Rink, Dieter. 2002. Environmental Policy and the Environmental Movement in Eastern Germany. *Capitalism, Nature, Socialism* 13 (3): 73–91.

Rip, Ari. 2004. Strategic Research, Post-modern Universities and Research Training. *Higher Education Policy* 17 (2): 153–166.

Robertson, David P. 2002. Public Ecology: Linking People, Science and the Environment. Paper presented at the Socially Robust Ecological Design Workshop, Bielefeld, Germany, October.

Robertson, David P., and R. Bruce Hull. 2003. Public Ecology: An Environmental Science and Policy for Global Society. *Environmental Science and Policy* 6 (5): 399–410.

Rohrbach, Daniela. 2007. The Development of Knowledge Societies in Nineteen OECD Countries between 1970 and 2002. *Social Sciences Information. Information sur les sciences sociales* 46 (4): 655–689.

Rosa, Eugene A. 2003. The Logical Structure of the Social Amplification of Risk Framework (SARF). In *Risk Communication and Social Amplification of Risk*, ed. Nick Pidgeon, Roger Kasperson, and Paul Slovic, 46–76. Cambridge: Cambridge University Press.

Rosa, Eugene, and Lauren Richter. 2008. Durkheim on the Environment: Ex Libris or Ex Cathedra? *Organization and Environment* 21 (2): 182–187.

Rosa, Hartmut. 2005. *Beschleunigung: Die Veränderung der Zeitstruktur in der Moderne.* Frankfurt am Main: Suhrkamp.

Ross, Edward A. 1896. Uncertainty as a Factor in Production. *Annals of the American Academy of Political and Social Science* 8 (2): 92–119.

Ross, Edward A. [1901] 2009. *Social Control: A Survey of the Foundations of Order.* New York: Transaction.

Ross, Laurel M. 1997. The Chicago Wilderness and Its Critics. 1, The Other Side: A Coalition for Urban Conservation. *Restoration and Management Notes* 15 (1): 17–24.

Rule, James B., and Yasemin Besen. 2008. The Once and Future Information Society. *Theory and Society* 37 (4): 317–342.

Ryle, Gilbert. 1949. *The Concept of Mind.* London: Hutchinson.

Saloma-Akpedonu, Czarina. 2008. Malaysian Technological Elite: Specifics of a Knowledge Society in a Developing Country. *Perspectives on Global Development and Technology* 7 (1): 1–14.

Salter, Liora. 1988. *Mandated Science: Science and Scientists in the Making of Standards.* Dordrecht: Kluwer.

Sarewitz, Daniel, Roger A. Pielke, Jr., and Radford Byerly, Jr., eds. 2000. *Prediction: Science, Decision Making, and the Future of Nature.* Washington, DC: Island Press.

Saunders, Peter T. 2000. Use and Abuse of the Precautionary Principle. London, Institute for Science and Society. http://www.i-sis.org.uk/prec.php (last accessed on March 20, 2008).

Schäfer, Wolf, ed. 1983. *Finalization of Science: The Social Orientation of Scientific Progress.* Dordrecht: Reidel.

Schatzki, Theodore R. 1996. *Social Practices: A Wittgensteinian Approach to Human Activity and the Social.* Cambridge: Cambridge University Press.

Scheler, Max. 1926. *Die Wissensformen und die Gesellschaft.* Leipzig: Neue-Geist-Verlag.

Schlenstedt, Jörg, and Joachim Bender. 2004. Programme of the LMBV-mbH to Protect Areas in Post-mining Landscapes with High Relevance for Conservation. *Peckiana* 3:113–117.

Schneider, Louis. 1962. The Role of the Category of Ignorance in Sociological Theory: An Exploratory Statement. *American Sociological Review* 27 (4): 492–508.

Schneider, Stephen H. 2001. Environmental Surprise. In *International Encyclopedia of the Social and Behavioral Sciences*, ed. Neil J. Smelser and Paul B. Baltes, 4671–4675. Oxford: Elsevier.

Scholz, Roland W., Daniel J. Lang, Arnim Wiek, Alexander I. Walter, and Michael Stauffacher. 2006. Transdisciplinary Case Studies as a Means of Sustainability Learning: Historical Framework and Theory. *International Journal of Sustainability in Higher Education* 7 (3):226–251.

Schreck, Peter. 1998. Environmental Impact of Uncontrolled Waste Disposal in Mining and Industrial Areas in Central Germany. *Environmental Geology* 35 (1): 66–72.

Schultze, Martin, Kurt Friese, René Frömmichen, Walter Geller, Helmut Klapper, and Katrin Wendt-Potthoff. 1999. Tagebaurestseen: Schon bei der Entstehung ein Sanierungsfall. *Gaia* 8 (1): 32–43.

Schulz, Winfried. 1970. *Kausalität und Experiment in den Sozialwissenschaften.* Mainz: Hase & Koehler.

Schwartz, Peter. 2003. *Inevitable Surprises: Thinking Ahead in a Time of Turbulence.* New York: Gotham Books.

Science and Policy Working Group. 2005. The SER Primer on Ecological Restoration. http://www.ser.org.

Sellke, Piet, and Ortwin Renn. 2010. Risk, Society and Environmental Policy: Risk Governance in a Complex World. In *Environmental Sociology: European Perspectives and Interdisciplinary Challenges*, ed. Matthias Gross and Harald Heinrichs. Berlin: Springer.

Seyfang, Gill. 2009. *The New Economics of Sustainable Consumption: Seeds of Change.* Houndmills: Palgrave Macmillan.

Shackle, George L. S. [1972] 1991. *Epistemics and Economics: A Critique of Economic Doctrines.* New York: Transaction.

Shackley, Simon, and Brian Wynne. 1996. Representing Uncertainty in Global Climate Change Science and Policy: Boundary-Ordering Devices and Authority. *Science, Technology and Human Values* 21 (3): 275–302.

Shalin, Dmitri. 1992. Critical Theory and the Pragmatist Challenge. *American Journal of Sociology* 98 (2): 237–279.

Shapin, Steven, and Simon Schaffer. 1985. *Leviathan and the Air-Pump: Hobbes, Boyle, and the Experimental Life.* Princeton, NJ: Princeton University Press.

Shinn, Terry. 2005. New Sources of Radical Innovation: Research-Technologies, Transversality and Distributed Learning in a Post-Industrial Order. *Social Sciences Information. Information sur les sciences sociales* 44 (4): 731–764.

Shnayerson, Michael. 2008. *Coal River.* New York: Farrar, Straus & Giroux.

Shrum, Wesley, Joel Genuth, and Ivan Chompalow. 2007. *Structures of Scientific Collaboration.* Cambridge, MA: MIT Press.

Siebel, Wigand. 1965. *Die Logik des Experiments in den Sozialwissenschaften.* Berlin: Duncker & Humblot.

Siegrist, Michael, Tomothy C. Earle, and Heinz Gutscher, eds. 2007. *Trust in Cooperative Risk Management: Uncertainty and Scepticism in the Public Mind.* London: Earthscan.

Siewers, Alf. 1998. Making the Quantum-Culture Leap: Reflections on the Chicago Controversy. *Restoration and Management Notes* 16 (1): 9–15.

Simmel, Georg. 1906. The Sociology of Secrecy and of Secret Societies. *American Journal of Sociology* 11 (4): 441–498.

Simmel, Georg. [1908] 1992. *Soziologie: Untersuchungen über die Formen der Vergesellschaftung.* Frankfurt am Main: Suhrkamp.

Simmel, Georg. [1911] 1998. *Philosophische Kultur. Gesammelte Essais.* Berlin: Wagenbach.

Simmel, Georg. [1918] 1999. Lebensanschauung. In *Der Krieg und die geistigen Entscheidungen et al., Gesamtausgabe 16*, ed. Gregor Fitzi and Otthein Rammstedt, 297–345. Frankfurt am Main: Suhrkamp.

Simmel, Georg. 1922. Über Freiheit. *Logos: Internationale Zeitschrift für Philosophie der Kultur* 11 (1):1–30.

Simmel, Georg. 1964. *The Sociology of Georg Simmel*. Wolff, Kurt H., (trans.). Glencoe, IL: Free Press.

Simon, Herbert A. 1957. *Models of Man: Social and Rational*. New York: Wileys.

Sitkin, Sim B. 1995. Learning through Failure: The Strategy of Small Losses. In *Organizational Learning*, ed. Michael D. Cohen and Lee S. Sproull, 541–577. Thousand Oaks, CA: Sage.

Small, Albion W., and George E. Vincent. 1894. *An Introduction to the Science of Society*. New York: American Book Co.

Smith, Julia L. 2008. A Critical Appreciation of the "Bottom-up" Approach to Sustainable Water Management: Embracing Complexity Rather than Desirability. *Local Environment* 13 (4): 353–366.

Smithson, Michael. 1989. *Ignorance and Uncertainty: Emerging Paradigms*. New York: Springer.

Smithson, Michael. 1990. Ignorance and Disasters. *International Journal of Mass Emergencies and Disasters* 8 (3): 207–235.

Smithson, Michael. 1993. Ignorance and Science: Dilemmas, Perspectives, and Prospects. *Knowledge: Creation, Diffusion, Utilization* 15 (2): 133–156.

Smithson, Michael. 2008. Social Theories of Ignorance. In *Agnotology: The Making and Unmaking of Ignorance*, ed. Robert N. Proctor and Londa Schiebinger, 209–229. Stanford: Stanford University Press.

Spaargaren, Gert, Arthur P. J. Mol, and Frederick H. Buttel, eds. 2000. *Environment and Global Modernity*. London: Sage.

Spencer, Herbert. [1850] 1970. *Social Statics: Or, the Conditions Essential to Human Happiness Specified, and the First of Them Developed*. Farnborough: Gregg.

Stehr, Nico. 1994. *Knowledge Societies: The Transformation of Labor, Property and Knowledge in Contemporary Societies*. London: Sage.

Stehr, Nico. 2001. *The Fragility of Modern Societies: Knowledge and Risk in the Information Age*. London: Sage.

Stehr, Nico. 2006. Knowledge and the Knowledge Society. In *Encyclopedia of Social Theory*, ed. Austin Harrington, Barbara L. Marshall, and Hans-Peter Müller, 305–307. London: Routledge.

Stichweh, Rudolf. 2004. Wissensgesellschaft und Wissenschaftssystem. *Swiss Journal of Sociology* 30 (2): 147–165.

Stocking, S. Holly. 1998. On Drawing Attention to Ignorance. *Science Communication* 20 (1): 165–178.

Stocking, S. Holly, and Lisa W. Holstein. 1993. Constructing and Reconstructing Scientific Ignorance. *Knowledge: Creation, Diffusion, Utilization* 15 (2): 186–210.

Stoepel, Beatrix. 2004. *Wölfe in Deutschland*. Hamburg: Hoffmann & Campe.

Strauch, Gerhard, and Walter Glässler. 1998. Kippen und Tagebaurestseen: Ein Problem für die Nachnutzung. *Geospektrum* 1 (5): 9–11.

Strulik, Torsten. 2004. *Nichtwissen und Vertrauen in der Wissensökonomie*. Frankfurt am Main: Campus.

Sunstein, Cass R. 2005. *Laws of Fear: Beyond the Precautionary Principle*. Cambridge: Cambridge University Press.

Svensson, Lennart G. 2006. New Professionalism, Trust and Competence: Some Conceptual Remarks and Empirical Data. *Current Sociology* 54 (4): 579–593.

Swart, Jac A. A., and Jelte van Andel. 2008. Rethinking the Interface between Ecology and Society: The Case of the Cockle Controversy in the Dutch Wadden Sea. *Journal of Applied Ecology* 45 (1): 82–90.

Szerszynski, Bronislaw. 2003. Technology, Performance and Life Itself: Hannah Arendt and the Fate of Nature. In *Nature Performed: Environment, Culture and Performance*, ed. Bronislaw Szerszynski, Wallace Heim, and Claire Waterton, 203–218. Oxford: Blackwell.

Tacke, Veronika. 2001. BSE as an Organizational Construction: A Case Study on the Globalization of Risk. *British Journal of Sociology* 52 (2): 293–312.

Taleb, Nassim N. 2007. *The Black Swan: The Impact of the Highly Improbable*. New York: Random House.

Tannert, Christof, Horst-Dietrich Elvers, and Burkhard Jandrig. 2007. The Ethics of Uncertainty. *EMBO Reports* 8 (10): 892–896.

Taupitz, Jochen. 1998. Das Recht auf Nichtwissen. In *Festschrift für Günther Wiese zum 70. Geburtstag*, ed. Peter Hanau, Egon Lorenz, and Hans-Christoph Matthes, 538–602. Neuwied: Luchterhand.

Taylor-Gooby, Peter, and Jens O. Zinn, eds. 2006. *Risk in Social Science*. Oxford: Oxford University Press.

Temperton, Vicky M. 2007. The Recent Double Paradigm Shift in Restoration Ecology. *Restoration Ecology* 15 (2): 344–347.

Ter Vehn, Jörg. 2006. Kalk soll für gutes Wasser sorgen. *Leipziger Volkszeitung (LVZ)*, January 26: 19.

Terwel, Bart W., Fieke Harinck, Naomi Ellemers, and Dancker D. L. Daamen. 2009. How Organizational Motives and Communications Affect Public Trust in Organizations: The Case of Carbon Dioxide Capture and Storage. *Journal of Environmental Psychology* 29 (2): 290–299.

Thomas, William I., and Dorothy Swaine Thomas. 1928. *The Child in America: Behavior Problems and Programs*. New York: Knopf.

Thompson, Michael. 1986. Commentary. In *Sustainable Development of the Biosphere*, ed. William C. Clark and R. E. Munn, 453–454. Cambridge: Cambridge University Press.

Thompson-Klein, Julie, Walter Grossenbacher-Mansuy, Rudolf Häberli, Alain Bill, Roland W. Scholz, and Myrtha Welti, eds. 2001. *Transdisciplinarity: Joint Problem Solving among Science, Technology, and Society*. Basel: Birkhäuser.

Tiefensee, Wolfgang. 2005. Geleitwort: Leipzig, eine Stadt im Aufbruch. In *Der Leipzig Atlas: Unterwegs in einer weltoffenen Stadt am Knotenpunkt zwischen West und Osteuropa*, ed. Helga Schmidt, 6. Cologne: Emons Verlag.

Timmerman, Peter. 1986. Mythology and Surprise in the Sustainable Development of the Biosphere. In *Sustainable Development of the Biosphere*, ed. William C. Clark and R. E. Munn, 435–453. Cambridge: Cambridge University Press.

Tischew, Sabine, and Anita Kirmer. 2007. Implementation of Basic Studies in the Ecological Restoration of Surface-Mined Land. *Restoration Ecology* 15 (2): 321–325.

Tolman, Frank L. 1902. The Study of Sociology in Institutions of Learning in the United States (Part 2). *American Journal of Sociology* 8 (1): 85–121.

Tremmel, Jörg. 2005. *Bevölkerungspolitik im Kontext ökologischer Generationengerechtigkeit*. Wiesbaden: Deutscher Universitätsverlag.

Turner, Frederick. 1987. The Self-Effacing Art: Restoration as Imitation of Nature. In *Ecological Restoration: A Synthetic Approach to Ecological Research*, ed. William R. Jordan III, Michael E. Gilpin, and John D. Aber, 47–52. Cambridge: Cambridge University Press.

Van Aken, Joan E. 2005. Management Research as a Design Science: Articulating the Research Products of Mode 2 Knowledge Production in Management. *British Journal of Management* 16 (1): 19–36.

Van Andel, Jelte, and James Aronson, eds. 2005. *Restoration Ecology*. London: Blackwell.

Van Creveld, Martin. 1999. *The Rise and Decline of the State*. Cambridge: Cambridge University Press.

Van den Daele, Wolfgang. 1981. Unbeabsichtigte Folgen sozialen Handelns: Anmerkungen zur Karriere des Themas. In *Lebenswelt und soziale Probleme,* ed. Joachim Matthes, 237–245. Frankfurt am Main: Campus Verlag.

Van der Ryn, Sim, and Stuart Cowan. 1996. *Ecological Design.* Washington, DC: Island Press.

Van der Windt, Henny J., and Jac A. A. Swart. 2008. Ecological Corridors, Connecting Science and Politics: The Case of the Green River in the Netherlands. *Journal of Applied Ecology* 45 (1): 124–132.

Vaughan, Diane. 1996. *The* Challenger *Launch Decision: Risky Technology, Culture, and Deviance at NASA.* Chicago: University of Chicago Press.

Veblen, Thorstein. 1898. Why Is Economics Not an Evolutionary Science? *Quarterly Journal of Economics* 12 (4): 373–397.

Wallerstein, Immanuel. 2004. *The Uncertainties of Knowledge.* Philadelphia: Temple University Press.

Walters, Carl J. 1986. *Adaptive Management of Renewable Resources.* New York: Macmillan.

Walters, Carl J., and Crawford S. Holling. 1990. Large-Scale Management Experiments and Learning by Doing. *Ecology* 71 (6): 2060–2068.

Wattenberg, Ben J. 2004. *Fewer: How the New Demography of Depopulation Will Shape Our Future.* Chicago: Dee.

Wätzold, Frank. 2000. Efficiency and Applicability of Economic Concepts Dealing with Environmental Risk and Ignorance. *Ecological Economics* 33 (2): 299–311.

Weber, Max. [1919] 1992. *Politik als Beruf.* Stuttgart: Reclam.

Wehling, Peter. 2006. *Im Schatten des Wissens? Perspektiven der Soziologie des Nichtwissens.* Konstanz: Universitätsverlag Konstanz.

Weick, Karl E. 2001. *Making Sense of the Organization.* Malden, MA: Blackwell.

Weick, Karl E., and Kathleen M. Sutcliffe. 2007. *Managing the Unexpected: Resilient Performance in an Age of Uncertainty.* New York: Wiley.

Weinberg, Alvin. 1972. Trans-Science. *Minerva* 10 (2): 209–222.

Weingart, Peter. 1997. From "Finalization" to "Mode 2": Old Wine in New Bottles? *Social Sciences Information. Information sur les sciences sociales* 36 (4): 591–613.

Weingart, Peter. 2008. How Robust Is "Socially Robust Knowledge." In *The Challenge of the Social and the Pressure of Practice: Science and Values Revisited,* ed. Martin Carrier, Don Howard, and Jeff Kourany, 131–145. Pittsburgh: University of Pittsburgh Press.

Weingart, Peter, Martin Carrier, and Wolfgang Krohn. 2007. *Nachrichten aus der Wissensgesellschaft: Analysen zur Veränderung der Wissenschaft.* Weilerswirst: Vebrück.

Weinstein, Michael P. 2007. Linking Restoration Ecology and Ecological Restoration in Estuarine Landscapes. *Estuaries and Coasts* 30 (2): 365–370.

Weinstein, Michael P. 2008. Ecological Restoration and Estuarine Management: Placing People in the Coastal Landscape. *Journal of Applied Ecology* 45 (1): 296–304.

Weinstein, Deena, and Michael A. Weinstein. 1978. The Sociology of Nonknowledge: A Paradigm. *Research in Sociology of Knowledge, Sciences and Art* 1: 151–166.

Wesselink, Anna J., Wiebe E. Bijker, Huib J. de Vriend, and Maarten S. Krol. 2007. Dutch Dealings with the Delta. *Nature and Culture* 2 (2): 188–209.

Westphal, Lynne. 2004. Restoring the Rustbelt: Social Science to Support Calumet's Ecological and Economic Revitalization. In *Integrated Resource and Environmental Management*, ed. Alan W. Ewert, Douglas C. Baker, and Glyn C. Bissix, 110–127. London: Cabi.

Westphal, Lynne M., Paul H. Gobster, and Matthias Gross. 2010. Models for Renaturing Brownfield Areas. In *Restoration and History: The Search for a Usable Environmental Past*, ed. Marcus Hall, 210–219. London: Routledge.

Whiteside, Kerry H. 2006. *Precautionary Politics: Principle and Practice in Confronting Environmental Risk.* Cambridge, MA: MIT Press.

Wibeck, Victoria. 2009. Communicating Uncertainty: Models of Communication and the Role of Science in Assessing Progress towards Environmental Objectives. *Journal of Environmental Policy and Planning* 11 (2):87–102.

Wildavsky, Aaron. 1988. *Searching for Safety.* New Brunswick: Transaction.

Wildavsky, Aaron. 1995. *But Is It True? A Citizen's Guide to Environmental Health and Safety Issues.* Cambridge, MA: Harvard University Press.

Williams, Stewart. 2008. Rethinking the Nature of Disaster: From Failed Instruments of Learning to a Post-Social Understanding. *Social Forces* 87 (2):1115–1138.

Willke, Helmut. 2002. *Dystopia: Studien zur Krisis des Wissens in der modernen Gesellschaft.* Frankfurt am Main: Suhrkamp.

Windelband, Wilhelm. 1980. [1894] History and Natural Science. *History and Theory* 19 (2): 169–185.

Winston, Andrew S., and Daniel J. Blais. 1996. What Counts as an Experiment? A Transdisciplinary Analysis of Textbooks, 1930–1970. *American Journal of Psychology* 109 (4): 599–616.

Wippler, Reinhard. 1978. Nicht-intendierte soziale Folgen individueller Handlungen. *Soziale Welt* 29 (2): 155–179.

Wolff, Kurt H. 1943. The Sociology of Knowledge: Emphasis on an Empirical Attitude. *Philosophy of Science* 10 (2): 104–123.

Wolff, Kurt H. 2002. *What It Contains*. Lanham, MD: Lexington Books.

Woodhouse, Edward J. 2007. The Lessons of Katrina for Intelligent Public Decision Making. *Nature and Culture* 2 (1): 10–26.

Woodhouse, Edward J., and David Collingridge. 1993. Incrementalism, Intelligent Trial-and-Error, and Political Decision Theory. In *An Heretical Heir of the Enlightenment*, ed. Harry Redner, 131–154. Boulder: Westview.

Wuthnow, Robert, and Wesley Shrum. 1983. Knowledge Workers as a "New Class": Structural and Ideological Convergence among Professional-Technical Workers and Managers. *Work and Occupations* 10 (4): 471–487.

Wynne, Brian. 1992. Uncertainty and Environmental Learning. *Global Environmental Change* 2 (2): 111–127.

Wynne, Brian. 2002. Some Observations on the Notion of "Experiment" Which Informs the Idea of Social Experiments for Robust Design. Paper presented at the Socially Robust Ecological Design Workshop, Bielefeld, Germany, October.

Young, Truman P. 2000. Restoration Ecology and Conservation Biology. *Biological Conservation* 92 (1): 73–83.

Zedler, Joy B. 2007. Success: An Unclear, Subjective Descriptor of Restoration Outcomes. *Ecological Restoration* 25 (3): 162–168.

Zerbe, Stefan, and Gerhard Wiegleb, eds. 2009. *Renaturierung von Ökosystemen in Mitteleuropa*. Heidelberg: Spektrum.

Zerubavel, Eviatar. 2006. *The Elephant in the Room: Silence and Denial in Everyday Life*. Oxford: Oxford University Press.

Ziman, John. 1996. Post-Academic Science: Constructing Knowledge with Networks and Norms. *Science Studies* 9 (1): 67–80.

Zimmermann, Ekkart. 1972. *Das Experiment in den Sozialwissenschaften*. Stuttgart: Teubner.

Znaniecki, Florian. 1940. *The Social Role of the Man of Knowledge*. New York: Columbia University Press.

Index

Printed in the United States
by Baker & Taylor Publisher Services

Printed in the United States
by Baker & Taylor Publisher Services